SPECIES-CHECKLIST OF THE BIRDS OF NEW GUINEA

Bruce M. Beehler
Brian W. Finch

ROYAL AUSTRALASIAN ORNITHOLOGISTS UNION MONOGRAPH No. 1
DECEMBER 1985

Royal Australasian Ornithologists Union,
21 Gladstone Street, Moonee Ponds, Victoria 3039.

ISSN 0815-2233
ISBN 0 9599823 2 9
RAOU Monograph No. 1

CONTENTS

SPECIES-CHECKLIST OF THE BIRDS OF NEW GUINEA

Bruce M. Beehler[1]
Brian W. Finch[2]

[1]NHB Room 336, Smithsonian Institution, Washington D.C. 20560, U.S.A.
[2]P.O. Box 496, Port Moresby, PNG.

INTRODUCTION

Published in 1941, Ernst Mayr's landmark *List of New Guinea Birds* has served for more than four decades as the key systematic reference to birds in the Papuan region. Not surprisingly, our knowledge of New Guinean birds has grown considerably since the appearance of that work. Some 750 publications that include data on New Guinean birds have appeared since 1940 (Frith 1979). Scores of taxonomic opinions have been argued on Papuan species, their generic position, and familial affiliation. It is gratifying to report that much in the fundamental outline, as well as the detail, of Mayr's pioneering volume remains unchallenged. Nonetheless, the need for a revised systematic species-list for the region's birds has grown year by year.

Austin L. Rand and E. Thomas Gilliard published the first one volume handbook to the birds of New Guinea in 1967. The taxonomic basis of their work was Mayr's checklist, although the Handbook introduced a number of nomenclatural and taxonomic departures from the original *List*. In their book, the use of English names gained in importance. Partly because of the influence of this book and Peckover & Filewood's (1976) popular book *Birds of New Guinea and Tropical Australia*, which highlighted additional name changes, two disparate English nomenclatures arose, one primarily 'American', the other 'Australian'. The student of Papuan birds who entered the field in the late 1970s found a confusing tangle of English and scientific names to unravel in the quest to learn the avifauna. An attempt to decipher Tom Iredale's iconoclastic *Birds of New Guinea* (1956) could have caused the beginner permanent taxonomic paralysis, what with the ungainly names and idiosyncratic opinions presented therein. Other publications often seemed to add to this nomenclatural logjam. There were simply *too many* names available for the birds of New Guinea, and no consensus as to which should be used to the exclusion of others.

In a collaborative effort with four other ornithologists specializing on the Papuan region, the present authors have recently completed a field guide to birds of New Guinea. As a foundation for that work, we felt it necessary to produce a new species-list of birds of the region, one that would incorporate the many relevant recent studies in systematics and nomenclature.

In doing this, we have deviated from Mayr's 1941 list in several ways. Whereas Mayr's list was distributional as well as taxonomic, ours omits data on distribution, for these are treated in the field guide (Beehler *et al.* 1986). Furthermore, in the present list we treat only to the species level. We do not present a subspecific analysis because such a task was beyond our present capacity; of necessity dealing with the details of many microgeographic distributions, such a task would require a number of years of museum study. We recognize the importance of such a project. On the one hand there is much uncertainty of subspecific nomenclature and on the other knowledge of subspecific distributions is only rudimentary for the Papuan region. Nevertheless, even though the present work is a species-list, we have been forced to assess well-marked allopatric populations (often considered as subspecies) in order to determine whether they should be given species-rank or not. Such decisions are often difficult and sometimes rather arbitrary. We have attempted to be as consistent as possible. These distinct populations, when treated as subspecies, are mentioned in the text firstly if there is some indication that they may merit specific rank, and secondly to highlight the need for further work on them.

Unlike Mayr's list, we include English names. We attempt to supply a fairly complete synonomy of the English epithets in recent usage (although Iredale's names have not been included). For wide-ranging species, we have made a special point to include differing English names that are in current use in Asia, Europe and elsewhere in the Pacific. Both old and new usages are included in the index to help the researcher quickly determine the present reference to lapsed names. Without an index, the job of determining the present status of a name in this list of more than 700 species would be a difficult process (see also Mordue & Finch 1982).

The Checklist

Geography

The following list treats the 708 species of birds documented as occurring in the Papuan region as of December 1983. As defined by Mayr (1941), the region comprises the entire island of New Guinea and its satellite islands: the Western and Eastern Papuan islands, the north coastal and Geelvink Bay islands, and Aru Islands. Note that the region *does not* include the islands of the Bismarck Archipelago (New Britain, New Ireland, etc.), the Solomons, the Moluccas, or any of the Australian reef islands in the Torres Strait (Boigu, Dauan, Saibai, and Deliverance islands are in Australian territory and are not treated here; see Draffan *et al.* 1983).

Treatment

The checklist is presented in phylogenetic sequence by orders, families and genera. We do not treat subfamilies, tribes, or subgenera in this sequence, and in some cases taxonomic groups that elsewhere are treated as subfamilies (cf. Orthonychinae, Turdinae, etc. in Peters' Checklist) are here listed with full familial status. We feel justified in doing this because the benefits of this hierarchical simplification far outweigh any loss of information (see Schodde 1975). This, in part, may be so because of the nature of the Papuan avifauna, and we are not recommending that this procedure be applied elsewhere without full consideration of the usefulness of the various subcategories.

For each species, we present the accepted scientific binomen with its original citation. This citation is useful for systematists, especially when conducting a taxonomic revision. In addition, the citation is valuable from an historical perspective. From these data we can see how long ago it was that a species was discovered and given a name, and it gives one the information needed to go to a library to read about past discoveries in Papuan ornithology, as told by the people who were intimately involved with that exciting process. We have not presented a technical synonomy. We believe this would be repetitive for such information is readily available in Mathews (1927-30), Mayr (1941), Peters' Checklist, Condon (1975), and other major faunal works.

We do provide references for taxonomic and nomenclatural opinion for all Papuan species. Most of these are found in revisions of orders or families, and citations appear in the introductory remarks for these sections. If some new, expanded (or restricted) use of the name is employed, then we note this. We are indebted to Richard Schodde, Murray D. Bruce, and Shane A. Parker for guidance and criticism on earlier drafts. George E. Watson provided essential advice in the process of researching and understanding original descriptions and in the complexities of the rules of zoological nomenclature. Additional authoritative opinion on taxonomic rules and nomenclature was provided by Curtis Sabrosky and George Steyskal. Finally, Leo Joseph and David Close were most patient, insightful and helpful in the editing and production of this work.

In many instances modern authorities disagree on the taxonomic definition of a species, and we have had to make a choice of which opinion to follow. In these situations, we cite both opinions, in order to allow the reader to follow-up on the taxonomic uncertainty.

English names

The process of selecting English names was time-consuming and often contentious. It involved weighing the opinions of a number of ornithological experts. Final decisions were made by a committee chaired by Brian Finch, and including Richard Schodde, Harry L. Bell, Dale A. Zimmerman, Thane K. Pratt, and Bruce Beehler. Additional valued opinions were solicited from Mary LeCroy, Jared M. Diamond, Ben King, and Walter E. Boles.

Alternate names were argued for and against, and in many instances votes were taken to gain a consensus. We based our judgements on six guidelines: (1) avoid coining a new name; (2) give priority to older names and ones that have the widest published usage; (3) for species with a primarily Australian distribution, conform to the most recently suggested nomenclature of the Royal Australasian Ornithologists Union's checklist committee (now in the process of revising the Australian list); (4) follow the guidelines implemented by King & Dickinson (1975) in their field guide to Southeast Asian birds; (5) give preference to brief names; and (6) choose a descriptive name over a patronym, unless the latter is significantly shorter or long-established. While these were guidelines, it seemed that nearly every species held some exceptional condition that made a final decision difficult. Almost every species has at least one English synonym. These are listed parenthetically in each species-account. Included also are variant spellings that have been widely used.

Checklist sequence

A quick perusal of the ordering of the passerine families in this list will show a radical departure from that in Mayr's (1941) *List* and Rand & Gilliard's (1967) *Handbook*. For the songbirds, those authors followed the familial sequence of Wetmore (1930, 1957; the 'finches last' sequence). Today, the almost universally accepted source of world systematic authority is the *Checklist of Birds of the World* begun in 1931 by James Lee Peters. Peters had originally intended to follow the Wetmore Sequence (Peters 1931: v), but subsequent to his death the *Checklist* tasks were assumed by a number of world authorities, and it was decided, by committee, that the sequence of passerines should follow the 'Basel Sequence' (with birds of paradise and crows last; Mayr & Greenway 1956).

We follow the Peters' Checklist sequence because New Guinea's closest biogeographical links are with Australia, and Australian authorities as well as all recent Australian ornithological works have followed this sequence. We attempt to bring some uniformity of treatment to the Australian zoogeographic realm. In addition, as Peters' *Checklist* nears completion, it gains in general acceptance, and it truly stands as a complete reference on the systematics of the world's birds. We view Peters' Checklist as the fundamental source and model for our list.

The avifauna

Our list of 708 species contains 59 more than that of Mayr (1941). The majority of these new records are of Palaearctic migrants and oceanic birds. Species new to science are few, but include the Long-bearded Melidectes *Melidectes princeps* (Mayr & Gilliard 1951).

Below we compare Mayr's avifaunal analysis with that in the present list.

	Mayr	*This List*
Breeding land birds:	568	578
Seabirds:	22	40
Palaearctic migrants:	37	56
Australian migrants:	22	34
TOTAL:	649	708

Verification

The requirements for including new taxa in the present list are not as formally codified as those constructed in the more populous and better-studied regions. We agree that, in most instances, a specimen or a series of excellent photographs best documents a new record. Nonetheless, we have accepted a number of well-documented sight records, and we have rejected others that, we feel, require further substantiation. If a species is morphologically distinctive, and if its world range is such that it is to be expected in the New Guinean region, then the likelihood of our accepting a sight record increases. Hypothetical records are included as notes within the text of related species, as well as in the Index.

We anticipate that this list will require periodic updating. Each new decade brings a measurable increase in the amount of ornithological research carried out in the New Guinean region. The pages of the *Papua New Guinea Bird Society Newsletter* annually document this growth in interest and knowledge. We would appreciate learning about new information on the Papuan avifauna, either in the form of submissions to the above *Newsletter* (c/– Editor, P.O. Box 1598, Boroko, Papua New Guinea) or as reprints or letters sent to the senior author of this work.

SYNOPSIS OF FAMILIES

Family	*Number of Species*
Casuariidae	3
Procellariidae	9
Oceanitidae	2
Podicipedidae	2
Phaethontidae	2
Fregatidae	2
Phalacrocoracidae	3
Anhingidae	1
Sulidae	3
Pelecanidae	1
Ardeidae	15
Ciconiidae	1
Threskiornithidae	4
Accipitridae	25
Falconidae	5
Anatidae	15
Megapodiidae	7
Phasianidae	3
Turnicidae	1
Rallidae	18
Gruidae	1
Otididae	1
Jacanidae	1
Haematopodidae	1
Recurvirostridae	1
Burhinidae	2
Glareolidae	2
Charadriidae	8
Scolopacidae	31
Phalaropodidae	1
Stercorariidae	3
Laridae	17
Columbidae	45
Psittacidae	46
Cuculidae	21
Tytonidae	4
Strigidae	6
Podargidae	2
Aegothelidae	6
Caprimulgidae	6
Hemiprocnidae	1
Apodidae	8
Alcedinidae	22
Meropidae	2
Coraciidae	1
Bucerotidae	1
Pittidae	3
Alaudidae	1
Hirundinidae	7
Motacillidae	4
Campephagidae	15
Laniidae	2
Turdidae	4
Orthonychidae	9
Pomatostomatidae	2
Sylviidae	9
Maluridae	5
Acanthizidae	20
Muscicapidae	1
Rhipiduridae	12
Myiagridae	20
Eopsaltriidae	25
Pachycephalidae	26
Climacteridae	1
Neosittidae	2
Dicaeidae	10
Nectariniidae	2
Zosteropidae	7
Meliphagidae	65
Estrildidae	16
Sturnidae	8
Oriolidae	4
Dicruridae	2
Grallinidae	2
Artamidae	3
Cracticidae	7
Ptilonorhynchidae	11
Paradisaeidae	38
Corvidae	3

SPECIES-CHECKLIST

ORDER STRUTHIONIFORMES

Ordinal designations and sequence from the Struthioniformes through the Falconiformes follow Mayr & Cottrell (1979).

FAMILY **CASUARIIDAE:** CASSOWARIES

Treatment follows Mayr (1979b).

001. **Casuarius bennetti** Dwarf Cassowary
Casuarius bennetti Gould, 1857 (1858), Proc. Zool. Soc. Lond.: 269, pl. 129.
(Little or Mountain Cassowary, *C. bennettii*)

002. **Casuarius casuarius** Southern Cassowary
Struthio Casuarius Linnaeus, 1758, Syst. Nat., ed. 10, 1: 155.
(Double-wattled, Two-wattled, or Australian Cassowary)

003. **Casuarius unappendiculatus** Northern Cassowary
Casuarius unappendiculatus Blyth, 1860, Journ. Asiat. Soc. Bengal 29: 112.
(Single-wattled or One-wattled Cassowary)

ORDER PROCELLARIIFORMES

Treatment follows Jouanin & Mougin (1979).

FAMILY **PROCELLARIIDAE:** PETRELS, SHEARWATERS AND ALLIES

004. **Macronectes giganteus** Southern Giant Petrel
Procellaria gigantea Gmelin, 1789, Syst. Nat. 1: 563.
A bird banded on South Georgia was recovered in Papua New Guinea (Hudson & Conroy 1975).
(Antarctic Giant-Petrel, Giant Fulmar)

005. **Pterodroma rostrata** Tahiti Petrel
Procellaria rostrata Peale, 1848, U.S. Explorer Exped. 8: 296.
First recorded in the New Guinean region by McKean & Vernon (1971). Note: *Pterodroma neglecta* and *P. leucoptera* may occur in the region as vagrants.

006. **Pachyptila turtur** Fairy Prion
Procellaria turtur Kuhl, 1820, Beitr. Zool. Vergl. Anat., Abth. 1: 143.

007. **Calonectris leucomelas** Streaked Shearwater
Procellaria leucomelas Temminck, 1835, Pl. Col., livr. 99: pl. 587.
(*Puffinus leucomelas*, White-faced or Streak-headed Shearwater)

008. **Puffinus pacificus** Wedge-tailed Shearwater
Procellaria pacifica Gmelin, 1789, Syst. Nat. 1: 560.

009. **Puffinus carneipes** Flesh-footed Shearwater
Puffinus carneipes Gould, 1844, Ann. Mag. Nat. Hist. 13: 365.
Recorded in region by Coates (1970a).
(Fleshy-footed or Pink-footed Shearwater)

010. **Puffinus griseus** Sooty Shearwater
Procellaria grisea Gmelin, 1789, Syst. Nat. 1: 564.
First recorded in the New Guinean region by Greensmith (1975).

011. **Puffinus tenuirostris** Short-tailed Shearwater
Procellaria tenuirostris Temminck, 1835, Pl. Col. livr. 99: text to plate 587.
Recorded by H. L. Bell (pers. comm.): sight records and specimens.
(Bonaparte's or Slender-billed Shearwater)

012. **Puffinus nativitatis** Christmas Shearwater
Puffinus (Nectris) nativitatis Streets, 1877, Bull. U.S. Nat. Mus. 7: 29.
Recorded by Finch (1981b). Note: *Puffinus lherminieri* may be a vagrant to region.
(Christmas Island Shearwater)

FAMILY **OCEANITIDAE:** STORM-PETRELS

Brodkorb (1963: 246) has shown that the name Hydrobatidae is preoccupied.

013. **Oceanites oceanicus** Wilson's Storm-Petrel

Procellaria oceanica Kuhl, 1820, Beitr. Zool. Vergl. Anat., Abth. 1: 136, pl. 10, fig. 1.

Sight records were made off southern New Guinea by Hoogerwerf (1964).

014. **Fregetta grallaria** White-bellied Storm-Petrel

Procellaria grallaria Vieillot, 1817, Nouv. Dict. Hist. Nat., nouv. éd. 25: 418.

F. tropica may occur in the region.

ORDER **PODICIPEDIFORMES**

FAMILY **PODICIPEDIDAE:** GREBES OR DABCHICKS

Treatment follows Storer (1979).

015. **Tachybaptus novaehollandiae** Australasian Grebe

Podiceps novae Hollandiae Stephens, 1826, in Shaw, General Zool. 13: 18.

Podiceps novaehollandiae sometimes has been treated as a race of *ruficollis* (e.g. Pizzey 1980).

(Australian Little Grebe, Australian or Black-throated Dabchick)

016. **Tachybaptus ruficollis** Little Grebe

Colymbus ruficollis Pallas, 1764, in Vroeg's Cat. Ois. Adumbr.: 6

(*Podiceps ruficollis*, Eurasian Little Grebe, Red-throated Little Grebe)

ORDER **PELECANIFORMES**

Treatment follows Dorst & Mougin (1979) with emendation.

FAMILY **PHAETHONTIDAE:** TROPICBIRDS

017. **Phaethon rubricauda** Red-tailed Tropicbird

Phaeton [sic] *rubricauda* Boddaert, 1783, Tabl. Pl. Enlum.: 57.

First recorded in New Guinean region by Tubb (1945).

018. **Phaethon lepturus** White-tailed Tropicbird

Phaeton [sic] *lepturus* Daudin, 1802, in Buffon, Hist. Nat., éd. Didot, Quadr. 14: 319.

(Long-tailed Tropicbird)

FAMILY **FREGATIDAE:** FRIGATEBIRDS

019. **Fregata minor** Great Frigatebird

Pelecanus minor Gmelin, 1789, Syst. Nat. 1: 572.

(Greater Frigatebird)

020. **Fregata ariel** Lesser Frigatebird

Atagen ariel G. R. Gray 1845, Gen. Birds 3:[669], col. pl. [185].

(Least Frigatebird)

FAMILY **PHALACROCORACIDAE:** CORMORANTS

021. **Phalacrocorax carbo** Great Cormorant

Pelecanus Carbo Linnaeus, 1758, Syst. Nat., ed. 10, 1: 133.

(Cormorant, Big Black, Large Black, or Black Cormorant)

022. **Phalacrocorax sulcirostris** Little Black Cormorant

Carbo sulcirostris Brandt, 1837, Bull. Sci. Acad. Imp. Sci. St.-Pétersbourg 3: col. 56.

023. **Phalacrocorax melanoleucos** Little Pied Cormorant

Hydrocorax melanoleucos Vieillot, 1817, Nouv. Dict. Hist. Nat., nouv. éd. 8: 88.

(*Halietor melanoleucos*)

FAMILY **ANHINGIDAE:** DARTERS

Considered by Dorst & Mougin (1979) to be a subfamily of the Phalacrocoracidae. We follow Mayr (1941) and Condon (1975) in giving this group familial status.

024. **Anhinga melanogaster** Darter

Anhinga melanogaster Pennant, 1769, Ind. Zool.: 13, pl. 12.

We follow Condon (1975) and Voous (1977) in specific treatment, but see also Mees (1982b: 12) for an additional view.

(*Anhinga rufa, A. novaehollanidae*, Australian Darter, Australasian or Oriental Darter, Australian Anhinga)

FAMILY **SULIDAE:** BOOBIES

025. **Sula dactylatra** Masked Booby
Sula dactylatra Lesson, 1831, Traité Ornith. 8: 601.
(Blue-faced Booby)

026. **Sula sula** Red-footed Booby
Pelecanus Sula Linnaeus, 1766, Syst. Nat., ed. 12, 1: 218.

027. **Sula leucogaster** Brown Booby
Pelecanus Leucogaster Boddaert, 1783, Table Pl. Enlum.: 57.

FAMILY **PELECANIDAE:** PELICANS

028. **Pelecanus conspicillatus** Australian Pelican
Pelecanus conspicillatus Temminck, 1824, Pl. Col. livr. 47: pl. 276.

ORDER **CICONIIFORMES**

FAMILY **ARDEIDAE:** HERONS, EGRETS AND ALLIES

Treatment follows Payne (1979) with emendation, but see also Bock (1958) and Hancock & Elliot (1978).

029. **Ardea pacifica** Pacific Heron
Ardea pacifica Latham, 1801, Index Ornith., Suppl.: 45.
(White-necked Heron)

030. **Ardea sumatrana** Great-billed Heron
Ardea sumatrana Raffles, 1822, Trans. Linn. Soc. London 13: 325.

031. **Egretta alba** Great Egret
Egretta alba Linnaeus, 1758, Syst. Nat., ed. 10, 1: 144.
Generic assignment follows Hancock (1984). Mock (1977) and van Tets (1977) recommend merging *Egretta* into *Ardea*.
(*Casmerodius albus, Ardea alba*, Large Egret, White Egret, Great White Egret, Great White Heron)

032. **Egretta picata** Pied Heron
Ardea (Herodias) picata Gould, 1845, Proc. Zool. Soc. London: 62.
(*Notophoyx picata, Hydranassa picata,* Pied Egret)

033. **Egretta intermedia** Intermediate Egret
Ardea intermedia Wagler, 1829, Isis von Oken: col. 659.
(*Mesophoyx intermedia*, Lesser, Plumed, or Yellow-billed Egret; Smaller or Short-billed Egret)

034. **Egretta ibis** Cattle Egret
Ardea ibis Linnaeus, 1758, Syst. Nat., ed. 10, 1: 144.
(*Bubulcus ibis, Ardeola ibis*)

035. **Egretta novaehollandiae** White-faced Heron
Ardea novae Hollandiae Latham, 1790, Index Ornith. 2: 701.
(*Notophoyx novehollandiae*)

036. **Egretta garzetta** Little Egret
Ardea Garzetta Linnaeus, 1766, Syst. Nat. ed. 12, 1: 237.
(Snowy Egret)

037. **Egretta sacra** Eastern Reef-Egret
Ardea sacra Gmelin, 1789, Syst. Nat. 1: 640.
(*Demigretta sacra*, Reef Heron, Eastern Reef Heron, Pacific Reef-Egret)

038. **Ardeola striata** Striated Heron
Ardea striata Linnaeus, 1758, Syst. Nat. ed. 10, 1: 144.
(*Butorides striatus*, Green, Mangrove or Green-backed Heron; Little Green or Little Heron)

039. **Nycticorax caledonicus** Rufous Night-Heron
Ardea caledonica Gmelin, 1789, Syst. Nat. 1: 626.
(Nankeen Night Heron, Night Heron)

040. **Zonerodius heliosylus** Forest Bittern
Ardea Heliosyla Lesson, 1828, Voy. Coquille, Atlas: pl. 44.
(Zebra Bittern or Zebra Heron)

041. **Ixobrychus minutus** Little Bittern
Ardea minuta Linnaeus, 1766, Syst. Nat. ed. 12, 1: 240.
(Black-backed Least Bittern)

042. **Ixobrychus sinensis** Yellow Bittern
Ardea Sinensis Gmelin, 1789, Syst. Nat. 1: 642.
(Chinese Least Bittern)

043. **Ixobrychus flavicollis** Black Bittern
Ardea flavicollis Latham, 1790, Index Ornith. 2: 701.
(Dupetor flavicollis)

FAMILY **CICONIIDAE:** STORKS

Treatment follows Kahl (1979).

044. **Ephippiorhynchus asiaticus** Black-necked Stork
Mycteria asiatica Latham, 1790, Index Ornith. 2: 670.
(*Xenorhynchus asiaticus*, Jabiru)

FAMILY **THRESKIORNITHIDAE:** IBISES AND SPOONBILLS

Treatment follows Steinbacher (1979) with emendation.

045. **Plegadis falcinellus** Glossy Ibis
Tantalus falcinellus Linnaeus, 1766, Syst. Nat., ed. 12, 1: 241.

046. **Threskiornis aethiopicus** Sacred Ibis
Tantalus aethiopicus Latham, 1790, Index Ornith. 2: 706.
Often considered specifically distinct from *aethiopicus* (e.g. Condon 1975).
(*Threskiornis molucca, T. moluccus*, White Ibis)

047. **Threskiornis spinicollis** Straw-necked Ibis
Ibis spinicollis Jameson, 1835, Edinburgh New Philos. Journ. 19: 213.

048. **Platalea regia** Royal Spoonbill
Platalea regia Linnaeus, 1758, Syst. Nat., ed. 10, 1: 139.
We follow Condon (1975) and others, in considering *regia* as specifically distinct from *leucorodia, contra* Steinbacher (1979).

ORDER **FALCONIFORMES**

Treatment follows Stresemann & Amadon (1979).

FAMILY **ACCIPITRIDAE:** HAWKS, EAGLES AND ALLIES

049. **Pandion haliaetus** Osprey
Falco Haliaetus Linnaeus, 1758, Syst. Nat. ed. 10, 1: 91.
Some treatments place this species in a monotypic family, the Pandionidae. Stresemann & Amadon (1979) consider that it constitutes a subfamily of the Accipitridae.
(Fish Hawk)

050. **Aviceda subcristata** Pacific Baza
Lepidogenys subcristatus Gould, 1838, Synops. Bds. Austr., pt. 3, pl. 46, and text.
(Crested Lizard Hawk, Crested Hawk or Baza)

051. **Henicopernis longicauda** Long-tailed Buzzard
Falco longicauda Garnot, 1828, Voy Coquille, Atlas: pl. 10; Zool. (1829) 1: 588.
(Long-tailed Honey-Buzzard)

052. **Macheiramphus alcinus** Bat Hawk
Macheiramphus alcinus Bonaparte, 1850, Rev. Mag. Zool., Paris; sér. 2, 2: 482.
Brooke & Clancy (1981) show that the older name of Bonaparte should be retained, *contra* Stresemann & Amadon (1979).
(*Machaerhamphus alcinus*)

053. **Elanus caeruleus** Black-shouldered Kite
Falco caeruleus Desfontaines, 1787 (1789), Hist. Acad. Roy. Sci., Paris: 503, pl. 15.
For notes on this species, known from the New Guinean region only since 1950, see Wood (1970) and Mees (1982b).
(Black-winged Kite)

054. **Milvus migrans** Black Kite
Falco migrans Boddaert, 1783, Table Pl. Enlum.: 28.
(Fork-tailed or Pariah Kite)

055. **Haliastur sphenurus** Whistling Kite

Milvus sphenurus Vieillot, 1818, Nouv. Dict. Hist. Nat., nouv. éd. 20: 564.

(Whistling Eagle)

056. **Haliastur indus** Brahminy Kite

Falco Indus Boddaert, 1783, Table Pl. Enlum.: 25.

(Red-backed Kite, Red-backed or White-headed Sea Eagle)

057. **Haliaeetus leucogaster** White-bellied Sea-Eagle

Falco leucogaster Gmelin, 1788, Syst. Nat. 1: 257.

(White-breasted Sea-Eagle)

058. **Circus spilonotus** Spotted Marsh-Harrier

Circus spilonotus Kaup, 1847, Isis von Oken: col. 953.

Brown & Amadon (1968) considered the two distinct New Guinean harrier populations to constitute two subspecies of the Old World harrier, *Circus aeruginosus*. We follow Mayr (1941), Stresemann & Amadon (1979) and others in keeping the Australasian populations distinct from *aeruginosus*. The status of these two populations is poorly understood. *C. spilonotus spilothorax* may be the sole breeding harrier population in New Guinea. Unconfirmed reports of *C. melanoleucus* from the highlands of New Guinea (cf. Crome & Swainson 1974) apparently refer to a previously undescribed plumage of *spilothorax*. More study of this group is needed.

(*Circus spilothorax*, Papuan Marsh Harrier, Spotted-backed Harrier)

059. **Circus approximans** Swamp Harrier

Circus approximans Peale, 1848, U.S. Explorer Exped. 8: 64.

Brown & Amadon (1968) include this population in the widespread *Circus aeruginosus*. Mees (1982b) showed that the Papuan birds are not subspecifically distinct from the Australian populations, and that records from New Guinea may be of non-breeding migrants. Baker-Gabb (1979) notes that only a single subspecies occurs throughout the Australasian region, supporting Mees's notion. Does *C. approximans* breed in New Guinea?

060. **Accipiter soloensis** Chinese Goshawk

Falco Soloensis Horsfield, 1821, Trans. Linn. Soc. London 13: 137.

(Horsfield's Goshawk, Frog Hawk)

061. **Accipiter fasciatus** Brown Goshawk

[Astur] fasciatus Vigors and Horsfield, 1827, Trans. Linn. Soc. London 15: 181.

(Australian Goshawk)

062. **Accipiter novaehollandiae** Grey Goshawk

Falco novae Hollandiae Gmelin, 1788, Syst. Nat. 1: 264.

(White Goshawk)

063. **Accipiter melanochlamys** Black-mantled Goshawk

Urospizias melanochlamys Salvadori, 1875, Ann. Mus. Civ. Stor. nat. Genova 7: 590.

064. **Accipiter poliocephalus** Grey-headed Goshawk

Accipiter poliocephalus G. R. Gray, 1858, Proc. Zool. Soc. London: 170.

(New Guinea Grey-headed Goshawk)

065. **Accipiter cirrhocephalus** Collared Sparrowhawk

Sparvius cirrhocephalus Vieillot, 1817, Nouv. Dict. Hist. Nat., nouv. éd. 10: 329.

066. **Accipiter meyerianus** Meyer's Goshawk

Astur Meyerianus Sharpe, 1878, Journ. Linn. Soc. London, Zool. 13: 458.

067. **Accipiter buergersi** Chestnut-shouldered Goshawk

Astur bürgersi Reichenow, 1914, Ornith. Monatsber. 22: 29.

(Buerger's or Burgers' Goshawk)

068. **Megatriorchis doriae** Doria's Hawk

Megatriorchis doriae Salvadori and D'Albertis, 1875, Ann. Mus. Civ. Stor. nat. Genova 7: 805.

Treatment follows Brown & Amadon (1968). Note that Amadon (1978) suggests placing this species in the genus *Accipiter*.

(Doria's Goshawk)

069. **Butastur indicus** Grey-faced Buzzard

Falco indicus Gmelin, 1788, Syst. Nat. 1: 264.

070. **Harpyopsis novaeguineae** New Guinea Harpy-Eagle

Harpyopsis novae guineae Salvadori, 1875, Ann. Mus. Civ. Stor. nat. Genova 7: 682.

(New Guinea Eagle, Kapul Eagle)

071. **Aquila gurneyi** Gurney's Eagle

Aquila (Heteropus?) gurneyi G. R. Gray, 1860, Proc. Zool. Soc. London: 342, pl. 169.

(*Spizaetus gurneyi*)

072. **Aquila audax** Wedge-tailed Eagle

Vultur audax Latham, 1801, Index Ornith., Suppl.: 2.

(*Uroaëtus audax*)

073. **Hieraaetus morphnoides** Little Eagle

Aquila morphnöides Gould, 1840 (1841), Proc. Zool. Soc. London: 161.

(Australian Little Eagle)

FAMILY **FALCONIDAE:** FALCONS

074. **Falco berigora** Brown Falcon

[Falco] berigora Vigors and Horsfield, 1827, Trans. Linn. Soc. London 15: 184.

(*Ieracidea berigora*, Brown Hawk)

075. **Falco cenchroides** Australian Kestrel

Falco cenchroides Vigors and Horsfield, 1827, Trans. Linn. Soc. London 15: 183.

(*Falco chenchroides* [sic], Nankeen Kestrel)

076. **Falco severus** Oriental Hobby

Falco severus Horsfield, 1821, Trans. Linn. Soc. London 13: 135.

077. **Falco longipennis** Australian Hobby

Falcon longipennis Swainson, 1837, Anim. Menag.: 341.

(Little Falcon)

078. **Falco peregrinus** Peregrine Falcon

Falco peregrinus Tunstall, 1771, Ornith. Brit., p. 1.

Note: There are sight records of *Falco hypoleucos* from southern and southeastern New Guinea (Stronoch 1981a, Finch 1981a).

ORDER **ANSERIFORMES**

Treatment follows Johnsgard (1979).

FAMILY **ANATIDAE:** DUCKS, GEESE, AND SWANS

079. **Anseranas semipalmata** Magpie Goose

Anas semipalmata Latham, 1798. Trans. Linn. Soc. London 4: 103.

(Pied or Semipalmated Goose)

080. **Dendrocygna guttata** Spotted Whistling Duck

Dendrocygna guttata Schlegel, 1866, Mus. Hist. Nat. Pays-Bas, Rev. Méthod. Crit. Coll. 6: 85.

(Spotted Tree Duck, Spotted Whistle-duck)

081. **Dendrocygna eytoni** Plumed Whistling Duck

Leptotarsis Eytoni Eyton, 1838, Monogr. Anatidae: 111.

(Plumed Tree-Duck, Grass Whistle-duck)

082. **Dendrocygna arcuata** Wandering Whistling Duck

Anas arcuata Horsfield, 1824, Zool. Researches Java 8: pl. 64.

(Water Tree-Duck, Water Whistle-Duck or Diving Whistling Duck)

083. **Cygnus atratus** Black Swan

Anas atrata Latham, 1790, Index Ornith.: 834.

First recorded in New Guinea by Beehler (1980).

084. **Tadorna radjah** White-headed Shelduck

Anas radjah 'Garnot' Lesson, 1828, Man. Ornith. 2: 417.

(Burdekin Duck, Rajah Shelduck)

085. **Nettapus pulchellus** Green Pygmy-Goose

Nettapus pulchellus Gould, 1842, Bds. Austr. 6: pl. 4.

086. **Nettapus coromandelianus** White Pygmy-Goose

Anas coromandeliana Gmelin, 1789, Syst. Nat. 1: 522.

(Cotton or White-quilled Pygmy-goose, Cotton Teal)

087. **Anas waigiuensis** Salvadori's Teal

Salvadorina waigiuensis Rothschild and Hartert, 1894, Novit. Zool. 1: 683.

Often placed in the monotypic genus *Salvadorina* (cf. Mayr 1941, Rand & Gilliard 1967).

(Salvadori's Duck)

088. **Anas penelope** Eurasian Wigeon

Anas penelope Linnaeus, 1758, Syst. Nat., ed. 10, 1: 126.

(Wigeon)

Note: there is a single unconfirmed report of *Anas castanea* from southern New Guinea (Stronoch 1980).

089. **Anas gibberifrons** Grey Teal

Anas (Mareca) gibberifrons S. Müller, 1842, in Temminck (ed.) Verh. Nat. Gesch., Land- en Volkenkunde: 159, note.

090. **Anas superciliosa** Pacific Black Duck

Anas superciliosa Gmelin, 1789, Syst. Nat. 1: 537.

Treated by some authorities as a race of *A. poecilorhyncha*.

(Black, Spot-billed, or Grey Duck)

091. **Anas acuta** Northern Pintail

Anas acuta Linnaeus, 1758, Syst. Nat., ed. 10, 1: 126.

First recorded in the region by Finch (1979a).

092. **Anas querquedula** Garganey

Anas Querquedula Linnaeus, 1758, Syst. Nat., ed. 10, 1: 126.

(Garganey Teal)

093. **Aythya australis** Australian White-eyed Duck

Nyroca australis Eyton, 1838, Monogr. Anatidae: 160.

(Hardhead, White-eyed Duck)

ORDER **GALLIFORMES**

FAMILY **MEGAPODIIDAE:** MEGAPODES OR MOUND-BUILDERS

This treatment follows Peters (1934), with emendations, as noted.

094. **Megapodius freycinet** Common Scrubfowl

Megapodius freycinet Gaimard, 1823, Bull. Gen. Univ. Anon. Nouv. Sci. 2: 451.

Our treatment for this species follows Mayr (1938) and Mees (1982b). We tentatively consider *freycinet* a polytypic, wide-ranging form in the New Guinean region. More study is needed of the geographic forms (cf. Schodde 1977, K. D. Bishop, pers. comm.).

(*Megapodius affinus, Megapodius reinwardt*, Common Scrub Hen, Scrub Fowl, Megapode, Incubator Bird)

095. **Megapodius wallacei** Moluccan Scrubfowl

Megapodius wallacei G. R. Gray, 1860, Proc. Zool. Soc. London: 362, pl. 171.

We follow the generic treatment of Ripley (1964b).

(*Eulipoa wallacei*, Moluccas Scrub Hen, Wallace's Scrubfowl, Moluccan Megapode)

096. **Talegalla cuvieri** Red-billed Brush-turkey

Talegalle Cuvieri Lesson, 1828, Man. Ornith. 2: 186.

(Red-billed Scrubturkey)

097. **Talegalla fuscirostris** Black-billed Brush-turkey

Talegallus fuscirostris Salvadori, 1877, Ann. Mus. Civ. Stor. nat. Genova Doria 9: 332.

(Black-billed Scrubturkey)

098. **Talegalla jobiensis** Brown-collared Brush-turkey

Talegallus jobiensis A. B. Meyer, 1874, Sitzungsb. Akad. Wiss. Wien 69 (1): 74.

(Collared Scrubturkey)

099. **Aepypodius arfakianus** Wattled Brush-turkey

Talegallus arfakianus Salvadori, 1877, Ann. Mus. Civ. Stor. nat. Giacomo Doria 9: 333.

(Wattled Scrubturkey)

100. **Aepypodius bruijnii** Waigeo Brush-turkey

Talegallus Bruijnii Oustalet, 1880, Compt. Rend. Acad. Sci., Paris 90: 906.

(Bruijn's Brush Turkey, Bruijn's Scrubturkey)

FAMILY **PHASIANIDAE:** QUAIL AND ALLIES

Treatment follows Condon (1975), and Peters (1934) for endemic Papuan taxa.

101. **Anurophasis monorthonyx** Snow Mountain Quail

Anurophasis monorthonyx van Oort, 1910, Notes Leyden Mus. 32: 212.

(*Coturnix monorthonyx*)

102. **Coturnix australis** Brown Quail

Perdix australis Latham, 1801, Index Ornith., Suppl.: 62.

(*Synoicus ypsilophorus, S. australis, Coturnix ypsilophorus*)

103. **Coturnix chinensis** King Quail

Tetrao chinensis Linnaeus, 1766, Syst. Nat. ed. 12, 1: 277.

(*Excalfactoria chinensis*, Chinese or Blue-breasted Quail)

ORDER **GRUIFORMES**

FAMILY **TURNICIDAE:** BUTTON-QUAIL

Treatment follows Condon (1975).

104. **Turnix maculosa** Red-backed Button-quail

Hemipodius maculosus Temminck, 1815, Pig. et Gall. 3: 631.

(Black-backed Bustard Quail, Red-backed Quail)

FAMILY **RALLIDAE:** RAILS, MOORHENS AND COOTS

Treatment follows Ripley (1977); however, for divergent views see also Olson (1973) and Mees (1982b).

105. **Eulabeornis castaneoventris** Chestnut Rail

Eulabeornis castaneoventris Gould, 1844, Proc. Zool. Soc. London: 56.

(Chestnut-bellied Rail)

106. **Eulabeornis plumbeiventris** Bare-eyed Rail

Rallus plumbeiventris G. R. Gray, 1861 (1862), Proc. Zool. Soc. London: 432.

(*Gymnocrex plumbeiventris*)

107. **Rallus philippensis** Buff-banded Rail

Rallus philippensis Linnaeus, 1766, Syst. Nat., ed. 12, 1: 263.

(*Gallirallus philippensis*, Banded Land Rail, Land Rail)

108. **Rallus torquatus** Barred Rail
Rallus torquatus Linnaeus, 1766, Syst. Nat., ed. 12, 1: 262.

109. **Rallus pectoralis** Lewin's Rail
Rallus pectoralis Temminck, 1831, Pl. Col. livr. 88: text to pl. 523, part.
(Water Rail, Slate-breasted Rail)

110. **Rallina rubra** Chestnut Forest-Rail
Rallicula rubra Schlegel, 1871 (1873), Nederl. Tijdschr. Dierk. 4: 55.
(Chestnut Rail)

111. **Rallina leucospila** White-striped Forest-Rail
Corethrura? leucospila Salvadori, 1875, Ann. Mus. Civ. Stor. nat. Genova Doria 7: 975.
(*Rallicula leucospila*, White-striped Rail, White-striped Chestnut Rail)

112. **Rallina forbesi** Forbes' Forest-Rail
Rallicula forbesi Sharpe, 1887, in Gould's Bds. of New Guinea 23: pl. [70] and text [=5 of bound volume].
(Forbes's Chestnut Rail, Forbes' Rail)

113. **Rallina mayri** Mayr's Forest-Rail
Rallicula rubra mayri Hartert, 1930, Novit. Zool. 36: 124.
See Diamond (1969) for a recent discussion of the taxonomy of this species and its allies.
(Mayr's Rail, Mayr's Chestnut Rail)

114. **Rallina tricolor** Red-necked Rail
Rallina tricolor G. R. Gray, 1858, Proc. Zool. Soc. London: 188.
(Red-necked Crake)

115. **Porzana cinerea** White-browed Crake
Porphyrio cinereus Vieillot, 1819, Nouv. Dict. Hist. Nat., nouv. éd. 28: 29.
(*Poliolimnas cinereus*, White-browed Rail)

116. **Porzana tabuensis** Spotless Crake
Rallus tabuensis Gmelin, 1789, Syst. Nat. 1: 717, no. 20.
(Sooty Rail)

117. **Porzana pusilla** Baillon's Crake
Rallus pusillus Pallas, 1776, Reise Versch. Prov. Russ. Reichs. 3: 700.
(Marsh or Lesser Spotted Crake)

118. **Amaurornis olivaceus** Bush-hen
Gallinula olivacea Meyen, 1834, Nova Acta Acad. Caes. Leop. Carol. 16, Suppl. 1: 109, pl. 20.
(*Gallinula olivacea*, Rufous-tailed Moorhen or Waterhen, Rufous-tailed Gallinule)

119. **Megacrex inepta** New Guinea Flightless Rail
Megacrex inepta D'Albertis and Salvadori, 1879, Ann. Mus. Civ. Stor. nat. Genova 14: 130.
We follow Mees (1982b) in retaining the genus *Megacrex*.
(*Habroptila inepta, Amaurornis ineptus*)

120. **Gallinula tenebrosa** Dusky Moorhen
Gallinula tenebrosa Gould, 1846, Bds. Austr. 22: pl. 14 (=6, pl. 73 of bound vol.)

121. **Porphyrio porphyrio** Purple Swamphen
Fulica Porphyrio Linnaeus, 1758, Syst. Nat., ed. 10, 1: 152.
(Swamphen)

122. **Fulica atra** Eurasian Coot
Fulica atra Linnaeus 1758, Syst. Nat., ed. 10, 1: 152.
(Coot, Common Coot, Black Coot)

FAMILY **GRUIDAE:** CRANES

Treatment follows Walkinshaw (1973).

123. **Grus rubicunda** Brolga
Ardea rubicunda Perry, 1810, Arcana 6: pl. 22.

The specific epithet *rubicunda* is adjectival, and must agree in gender with its feminine genus.

(*Grus rubicundus*, Australian Crane)

FAMILY **OTIDIDAE:** BUSTARDS

Treatment follows Condon (1975).

124. **Ardeotis australis** Australian Bustard

Otis australis J. E. Gray, 1829, Griffith's Anim. Kingd. 8: 305.

(*Choriotis australis, Eupodotis australis,* Bustard)

ORDER **CHARADRIIFORMES**

Familial sequence follows Jehl (1968). Nomenclature follows Condon (1975), with emendations and additions, as noted.

FAMILY **JACANIDAE:** JACANAS

125. **Irediparra gallinacea** Comb-crested Jacana

Parra gallinacea Temminck, 1828, Pl. Col., livr. 78: pl. 464.

(Lotusbird, Jacana, Lilytrotter)

FAMILY **HAEMATOPODIDAE:** OYSTERCATCHERS

126. **Haematopus longirostris** Pied Oystercatcher

Haematopus longirostris Vieillot, 1817, Nouv. Dict. Hist. Nat., nouv. éd. 15: 410.

Treatment follows Baker (1975) and McKean (1978), but see also Mees (1982b), who treats this form as a subspecies of *Haematopus ostralegus*.

FAMILY **RECURVIROSTRIDAE:** AVOCETS AND STILTS

127. **Himantopus leucocephalus** White-headed Stilt

Himantopus leucocephalus Gould, 1837, Synops. Bds. Austr. 2: pl. 34.

Treatment follows Mayr & Short (1970). In contrast, Condon (1975) treats this form as a subspecies of the worldwide *H. himantopus*.

(Black-winged or Pied Stilt)

FAMILY **BURHINIDAE:** STONE-CURLEWS AND THICK-KNEES

Treatment follows Schodde & Mason (1980).

128. **Burhinus grallarius** Bush Stone-curlew

Charadrius grallarius Latham, 1801, Index Ornith., Suppl.: 66.

First recorded in New Guinea by Lindgren (1971).

(*Burhinus magnirostris*, Bush Curlew, Southern Stone-curlew, Bush Thick-knee)

129. **Esacus magnirostris** Beach Stone-curlew

Oedicnemus magnirostris Vieillot, 1818, Nouv. Dict. Hist. Nat., nouv. éd. 23: 231.

(*Burhinus neglectus*, Beach Curlew, Great or Beach Thick-knee, Reef Stone Curlew, Great Stone Plover)

FAMILY **GLAREOLIDAE:** PRATINCOLES AND COURSERS

130. **Glareola maldivarum** Oriental Pratincole

Glareola (Pratincola) Maldivarum J. R. Forster, 1795, Faunula Indica, ed. 2: 11.

131. **Stiltia isabella** Australian Pratincole

Glareola isabella Vieillot, 1816, Analyse, p. 69.

(*Glareola isabella*, Australian Courser)

FAMILY **CHARADRIIDAE:** PLOVERS AND DOTTERELS

Treatment follows Condon (1975), and Peters (1934) for taxa not treated in Condon.

132. **Vanellus miles** Masked Lapwing

Tringa miles Boddaert, 1783, Table Pl. Enlum.: 51.

Cramp (1983) places this group into the genus *Hoplopterus*.

(*Lobibyx miles*, Masked Plover; includes *V. novaehollandiae*, Spur-winged Plover)

133. **Pluvialis squatarola** Grey Plover

Tringa squatrola Linnaeus, 1758, Syst. Nat. ed. 10, 1: 149.

(*Charadrius squatarola, Squatarola squatarola*, Black-bellied Plover)

134. **Pluvialis dominica**
Lesser Golden Plover

Charadrius Dominicus P. L. S. Müller, 1776, Natursyst., Suppl.: 116.

Conners (1983) considers the population *fulva*, which winters in New Guinea, to be specifically distinct from *dominica*.

(*Pluvialis fulva*, American, Eastern, or Pacific Golden Plover)

135. **Erythrogonys cinctus**
Red-kneed Dotterel

Erythrogonys cinctus Gould, 1838, Synops. Bds. Austr. 4: pl. 73.

First recorded in New Guinea by Hoogerwerf (1964).

(*Charadrius cinctus*)

136. **Charadrius dubius** Little Ringed Plover

Charadrius (dubius) Scolopi, 1786, Del. Flor. et Faun. Insubr. 2: 93.

137. **Charadrius mongolus**
Mongolian Plover

Charadrius mongolus Pallas, 1776, Reise Versch. Prov. Russ. Reichs. 3: 700.

(Lesser or Mongolian Sand-Plover, Lesser or Mongolian Sand Dotterel, Mongolian Dotterel)

Note: there are old, unconfirmed reports of *Charadrius (alexandrinus) ruficapillus* from New Guinea; no recent sightings.

138. **Charadrius leschenaultii**
Large Sand-Plover

Charadrius leschenaultii Lesson, 1826, Dict. Sci. Nat., éd. Levrault 42: 36.

(Large-billed or Large Dotterel, Greater Sand-Plover. Large Sand Dotterel)

139. **Charadrius veredus** Oriental Plover

Charadrius veredus Gould, 1848, Proc. Zool. Soc. London: 38.

Considered by some authorities to be conspecific with *Charadrius asiaticus*, but see Vaurie (1965).

(*Eupoda asiatica, E. veredus, Eupodia* [sic] *asiatica*)

FAMILY **SCOLOPACIDAE:** CURLEWS, SANDPIPERS, SNIPE AND ALLIES

Treatment follows Condon (1975), and Peters (1934) for those taxa not recorded from Australia.

140. **Arenaria interpres** Ruddy Turnstone

Tringa Interpres Linnaeus, 1758, Syst. Nat., ed. 10, 1: 148.

(Turnstone)

141. **Numenius madagascariensis** Eastern Curlew

Scolopax madagascariensis Linnaeus, 1766, Syst. Nat., ed. 12, 1: 242.

(Australian Curlew)

142. **Numenius phaeopus** Whimbrel

Scolopax Phaeopus Linnaeus, 1758, Syst. Nat. ed. 10, 1: 146.

143. **Numenius minutus** Little Curlew

Numenius minutus Gould, 1840 (1841), Proc. Zool. Soc. London: 176.

(Little Whimbrel)

144. **Tringa glareola** Wood Sandpiper

Tringa glareola Linnaeus, 1758, Syst. Nat., ed. 10, 1: 149.

145. **Tringa brevipes** Grey-tailed Tattler

Totanus brevipes Vieillot, 1816, Nouv. Dict. Hist. Nat., nouv. éd. 6: 410.

Earlier regional lists did not recognize the specific distinction of this species, lumping all populations into *incana*.

(*Heteroscelis brevipes*, Siberian Tattler)

146. **Tringa incana** Wandering Tattler

Scolopax incana Gmelin, 1789, Syst. Nat. 1 (2): 658.

Earlier treatments considered this and the previous species as conspecific, thus older reports of this species might actually refer to *brevipes*.

(*Heteroscelis incanus*)

147. **Tringa hypoleucos** Common Sandpiper

Tringa hypoleucos Linnaeus, 1758, Syst. Nat., ed. 10, 1: 149.

(*Actitis hypoleucos*)

148. **Tringa nebularia** Common Greenshank

Scolopax nebularia Gunnerus, 1767, in Leem's Beskr. Finm. Lapper: 251.

(Greenshank)

149. **Tringa totanus** Common Redshank

Scolopax totanus Linnaeus, 1758, Syst. Nat., ed. 10, 1: 145.

(Redshank)

First recorded in the region by Bell (1966a). See also Coates (1972).

150. **Tringa stagnatilis** Marsh Sandpiper

Totanus stagnatilis Bechstein, 1803, Orn. Taschenb. Deutschl.: 292, pl. 29.

First recorded in New Guinea by Hoogerwerf (1964).

(Little Greenshank)

151. **Tringa terek** Terek Sandpiper

Scolopax terek Latham, 1790, Index. Ornith. 2: 724.

(*Xenus cinereus*)

152. **Tringa ochropus** Green Sandpiper

Tringa Ocrophus Linnaeus, 1758, Syst. Nat., ed. 10, 1: 149, name subsequently emended to *ochropus*.

First recorded in region by Hoogerwerf (1964).

153. **Gallinago hardwickii** Latham's Snipe

Scolopax Hardwickii J. E. Gray, 1831, Zool. Misc. 1: 16.

(*Capella hardwickii*, Australian or Japanese Snipe)

154. **Gallinago megala** Swinhoe's Snipe

Gallinago megala Swinhoe, 1861, Ibis: 343.

(*Capella megala*, Marsh Snipe, or Chinese Snipe)

Note: *Gallinago stenura* has recently been reported from the Port Moresby area, but as this species is very difficult to identify in the field, we await the collection of a specimen before including it on this list.

155. **Scolopax saturata** Rufous Woodcock

Scolopax saturata Horsfield, 1821, Trans. Linn. Soc. London 13: 191.

(East Indian Woodcock)

156. **Limnodromus semipalmatus** Asian Dowitcher

Macrorhamphus semipalmatus 'Jerdon' Blyth, 1848, J. Asiat. Soc. Bengal 17: 252.

First recorded in the region by Heron (1977a).

(Asiatic or Snipe-billed Dowitcher)

157. **Limosa limosa** Black-tailed Godwit

Scolopax Limosa Linnaeus, 1758, Syst. Nat., ed. 10, 1: 147.

158. **Limosa lapponica** Bar-tailed Godwit

Scolopax lapponica Linnaeus, 1758, Syst. Nat., ed. 10, 1: 147.

159. **Calidris canutus** Red Knot

Tringa Canutus Linnaeus, 1758, Syst. Nat., ed. 10, 1: 149.

First recorded in region by Hoogerwerf (1964).

(Knot or Common Knot)

160. **Calidris tenuirostris** Great Knot

Totanus tenuirostris Horsfield, 1821, Trans. Linn. Soc. London 13: 192.

(Eastern Knot)

161. **Calidris acuminata** Sharp-tailed Sandpiper

Totanus acuminatus Horsfield, 1821, Trans. Linn. Soc. London 13: 192.

This and the following five species have been placed in the genus *Erolia* by some authorities (e.g. Peters 1934).

(*Erolia acuminata*, Siberian or Asian Pectoral Sandpiper)

162. **Calidris melanotos** Pectoral Sandpiper

Tringa melanotos Vieillot, 1819, Nouv. Dict. Hist. Nat., nouv. éd. 34: 462.

First recorded in the region by Coates (1973).

(*Erolia melanotos*)

163. **Calidris ruficollis** Red-necked Stint

Trynga ruficollis Pallas, 1776, Reise Versch. Prov. Russ. Reichs. 3: 700.

(*Erolia ruficollis*, Rufous-necked Stint, Eastern or Siberian Little Stint)

164. **Calidris minuta** Little Stint

Tringa minuta Leisler, 1812, Nachträge zu Bechstein's Naturg. Deutschl.: 74.

New Guinean records are detailed by Finch (1980a, 1983a).

(*Erolia minuta*)

165. **Calidris subminuta** Long-toed Stint

Tringa subminuta Middendorff, 1851, Reise Nord.-Ost. Sibir. 2: 222.

For New Guinean records see Heron (1978) and Finch (1979).

(*Erolia subminuta*, Middendorf's Stint)

166. **Calidris ferruginea** Curlew Sandpiper

Tringa Ferrugineus Pontoppidan, 1763, Danske Atlas 1: 624.

Use of the specific epithet *ferruginea* instead of *testacea* follows Stresemann (1941).

First recorded in the region by Hoogerwerf (1964).

(*Calidris testacea, Erolia ferruginea*)

167. **Calidris alba** Sanderling

Trynga alba Pallas, 1764, Vroeg's Cat. Ois. Adumbr.: 7.

(*Crocethia alba*)

168. **Tryngites subruficollis** Buff-breasted Sandpiper

Tringa subruficollis Vieillot, 1819, Nouv. Dict. Hist. Nat., nouv. éd. 34: 465.

First recorded in region by Mordue (1981).

169. **Limicola falcinellus** Broad-billed Sandpiper

Scolopax Falcinellus Pontoppidan, 1763, Danske Atlas 1: 623.

170. **Philomachus pugnax** Ruff

Tringa Pugnax Linnaeus, 1758, Syst. Nat. ed. 10, 1: 148.

First recorded in New Guinea by B. J. Coates (see Lindgren 1974).

(Reeve [=female])

FAMILY **PHALAROPODIDAE:** PHALAROPES

Treatment follows Condon (1975).

171. **Phalaropus lobatus** Red-necked Phalarope

Tringa tobata [sic] Linnaeus, 1758, Syst. Nat., ed. 10, 1: 148.

Several records from the seas north of Madang (Finch 1981b).

(*Lobipes lobatus*, Northern Phalarope)

FAMILY **STERCORARIIDAE:** SKUAS OR JAEGERS

Treatment follows Condon (1975).

172. **Stercorarius parasiticus** Arctic Skua

Larus parasiticus Linnaeus, 1758, Syst. Nat., Nat., ed. 10, 1: 136.

(Parasitic Jaeger)

First recorded in the region by Greensmith (1975). See also Finch (1981b).

173. **Stercorarius pomarinus** Pomarine Skua

Lestris pomarinus Temminck, 1815, Man. Ornith.: 514.

(Pomarine Jaeger)

174. **Stercorarius longicaudus** Long-tailed Skua

Stercorarius longicaudus Vieillot, 1819, Nouv. Dict. Hist. Nat., nouv. éd. 32: 157.

Condon (1975) altered the species name to *longicauda*, although the original usage by Vieillot has rendered the form of the specific name as an adjective (to agree in gender with the masculine genus). Recorded in the region by Greensmith (1975).

(*Stercorarius longicauda*)

FAMILY **LARIDAE:** GULLS AND TERNS

Treatment follows Condon (1975) and Peters (1934) for the species not recorded from Australia.

175. **Larus novaehollandiae** Silver Gull

Larus novaehollandiae Stephens, 1826, Shaw's Gen. Zool. 13 (1): 196.

First recorded in region by Bell (1966b).

(Red-billed Gull)

176. **Larus ridibundus** Common Black-headed Gull

Larus ridibundus Linnaeus, 1766, Syst. Nat., ed. 12, 1: 225.

First recorded in western New Guinea in 1963 by King (1979).

(Black-headed Gull)

177. **Chlidonias hybridus** Whiskered Tern

Sterna hybrida Pallas, 1811, Zoogr. Rosso.-Asiat. 2: 338.

The generic name is now considered to be masculine (ICZN).

(*C. hybrida*, Marsh Tern)

178. **Chlidonias leucopterus** White-winged Black Tern

Sterna leucoptera Temminck, 1815. Man. Ornith.: 483.

(*C. leucoptera*, White-winged Tern)

179. **Gelochelidon nilotica** Gull-billed Tern

Sterna nilotica Gmelin, 1789, Syst. Nat. 1: 606.

180. **Hydroprogne caspia** Caspian Tern

Sterna caspia Pallas, 1770, Novi Comment. Acad. Sci. Petropol. 14: 582, pl. 22, fig. 2.

181. **Sterna hirundo** Common Tern

Sterna hirundo Linnaeus, 1758, Syst. Nat. ed. 10, 1: 137.

(Eastern Common Tern)

182. **Sterna dougallii** Roseate Tern

Sterna Dougallii Montagu, 1813, Orn. Dict. Suppl.: text and plate to Tern, Roseate.

183. **Sterna sumatrana** Black-naped Tern

Sterna Sumatrana Raffles, 1822, Trans. Linn. Soc. London 13: 329.

184. **Sterna fuscata** Sooty Tern

Sterna fuscata Linnaeus, 1766, Syst. Nat. ed. 12, 1: 228.

185. **Sterna anaethetus** Bridled Tern

Sterna (anaethetus) Scopoli, 1786, Delic. Flor. Faun. Insubr. 2: 92.

(Brown-winged Tern)

186. **Sterna albifrons** Little Tern

Sterna (albifrons) Pallas, 1764, in Vroeg's Cat. Ois. Adumbr.: 6.

(Least Tern)

187. **Sterna bergii** Crested Tern

Sterna bergii Lichtenstein, 1823, Verz. Doubl. Zool. Mus. Univ. Berlin: 80.

(*Thalasseus bergii*, Great, Large, or Greater Crested Tern, Swift Tern)

188. **Sterna bengalensis** Lesser Crested Tern
Sterna bengalensis Lesson, 1831, Traité Ornith.: 621.
(Indian Lesser Crested Tern)

189. **Anous stolidus** Brown Noddy
Sterna stolida Linnaeus, 1758, Syst. Nat., ed. 10, 1: 137.
(Common Noddy)

190. **Anous minutus** Black Noddy
Anous minutus Boie, 1844, Isis von Oken 3: col. 188.
(White-capped Noddy)

191. **Gygis alba** White Tern
Sterna alba Sparrman, 1786, Mus. Carls. 1: no. 11.
(*Anous albus*, White Noddy, Fairy Tern)

ORDER **COLUMBIFORMES**

FAMILY **COLUMBIDAE**: PIGEONS AND DOVES

Treatment follows Goodwin (1970).

192. **Columba livia** Rock Pigeon
Columba domestica β livia Gmelin, 1789, Syst. Nat. 1 (2): 769.
(Rock Dove, Domestic or Feral Pigeon, Blue Rock Pigeon)

193. **Columba vitiensis**
White-throated Pigeon
Columba vitiensis Quoy and Gaimard, 1830, Voy. Astrolabe, Zool. 1: 246.

194. **Macropygia amboinensis**
Brown Cuckoo-Dove
Columba amboinensis Linnaeus, 1766, Syst, Nat. ed. 12, 1: 286.
(*Macropygia phasianella*, Amboina or Slender-billed Cuckoo-Dove, Brown Pigeon)

195. **Macropygia nigrirostris**
Black-billed Cuckoo-Dove
Macropygia nigrirostris Salvadori, 1875, Ann. Mus. Civ. Stor. nat. Genova 7: 972.
(*Macropygia nigriorostris* [sic], Rusty Cuckoo-dove)

196. **Macropygia mackinlayi**
Mackinlay's Cuckoo-Dove
Macropygia mackinlayi Ramsay, 1878, Proc. Linn. Soc. New South Wales 2: 286.
(Spot-breasted Cuckoo-Dove)

197. **Reinwardtoena reinwardtii**
Great Cuckoo-Dove
Columba reinwardtsi [sic] Temminck, 1824, Pl. Col. 42: 248; emended to *reinwardtii* by Temminck, 1838 (1839), Tabl. Méth.: 80.
(*Reinwartoena reinwardtsi*, Giant or Long-tailed Cuckoo-dove)

198. **Chalcophaps indica** Emerald Ground-Dove
Columba indica Linnaeus, 1758, Syst. Nat., ed. 10, 1: 164.
(Green-winged Pigeon, Green-winged Ground Dove, Emerald Dove)

199. **Chalcophaps stephani**
Stephan's Ground-Dove
Chalcophaps stephani Pucheran, 1853, Voy. Pôle Sud, Zool. 3: 119; Atlas, pl. 28, Fig. 2.
(Stephan's Green-winged Pigeon)

200. **Henicophaps albifrons** New Guinea Bronzewing
Henicophaps albifrons G. R. Gray, 1861 (1862), Proc. Zool. Soc. London: 432, pl. 44.
(White-capped Ground Pigeon, Jungle Bronzewing Pigeon)

201. **Geopelia striata** Peaceful Dove
Columba striata Linnaeus, 1766, Syst. Nat. ed. 12, 1: 282.
(*Geopelia placida*, Peaceful Ground Dove, Zebra Dove)

202. **Geopelia humeralis**
Bar-shouldered Dove

Columba humeralis Temminck, 1821, Trans. Linn. Soc. London 13 (1): 128.

203. **Caloenas nicobarica** Nicobar Pigeon
Columba nicobarica Linnaeus, 1758, Syst. Nat. ed. 10, 1: 164.

204. **Gallicolumba rufigula** Cinnamon Ground-Dove
Peristera rufigula Pucheran, 1853, Voy. Pôle Sud, Zool. 3: 118.
(Red-throated Ground Dove, Goldenheart)

205. **Gallicolumba beccarii** Bronze Ground-Dove
Chalcophaps beccarii Salvadori, 1875 (1876), Ann. Mus. Civ. Stor. nat. Genova 7: 974.
(Beccari's Ground Dove, Grey-bibbed Ground-dove)

206. **Gallicolumba jobiensis** White-bibbed Ground-Dove
Phlegoenas jobiensis A. B. Meyer, 1875, Mitt. Zool. Mus. Dresden 1: 10.
(White-throated Ground-dove)

207. **Trugon terrestris** Thick-billed Ground-Pigeon
Trugon terrestris G. R. Gray, 1849, Genera Bds. 3: appr. p. 24.
(Thick-billed Junglepigeon; Thick-billed Ground Dove)

208. **Otidiphaps nobilis** Pheasant Pigeon
Otidiphaps nobilis Gould, 1870, Ann. Mag. Nat. Hist., 4th ser., 5: 62.
(Magnificent or Noble Ground Pigeon, Magnificent Pheasant Pigeon)

209. **Goura cristata**
Western Crowned Pigeon
Columba cristata Pallas, 1764, in Vroeg's Cat. Ois., Adumbr.: 2.
(Common Crowned Pigeon, Masked Goura)

210. **Goura scheepmakeri**
Southern Crowned Pigeon
Goura scheepmakeri Finsch, 1875 (1876), Proc. Zool. Soc. London: 631, pl. 68.
(Sheepmaker's Crowned Pigeon, Great Goura or Guria)

211. **Goura victoria** Victoria Crowned Pigeon
Lophyrus victoria Fraser, 1844, Proc. Zool. Soc. London: 136.
(Victoria Goura or Guria)

212. **Ptilinopus magnificus**
Wompoo Fruit-Dove
Columba magnifica Temminck, 1821, Trans. Linn. Soc. London 13: 125.
(*Megaloprepia magnifica*, Wompoo Pigeon, Magnificent Fruitdove)

213. **Ptilinopus perlatus**
Pink-spotted Fruit-Dove
Columba perlata Temminck, 1835, Pl. Col., livr. 94: pl. 559.

214. **Ptilinopus ornatus** Ornate Fruit-Dove
Ptilinopus ornatus 'von Rosenberg' Schlegel, 1871 (1873), Nederl. Tijdschr. Dierk. 4: 52.

215. **Ptilinopus aurantiifrons**
Orange-fronted Fruit-Dove
Ptilonopus aurantiifrons G. R. Gray, 1858, Proc. Zool. Soc. London: 185, pl. 137.

216. **Ptilinopus wallacii**
Wallace's Fruit-Dove
Ptilonopus Wallacii G. R. Gray, 1858, Proc. Zool. Soc. London: 185, pl. 136.
(Wallace's Green Fruit Dove)

217. **Ptilinopus superbus**Superb Fruit-Dove
Columba Superba Temminck, 1810, in Knip, Les Pigeons, les Colombes: 75, pl. 33.
(Purple-crowned Pigeon)

218. **Ptilinopus regina**
Rose-crowned Fruit-Dove

Ptilinopus purpuratus var. *Regina* Swainson, 1825, Zool. Journ. 1: 474.

(Swainson's Fruitdove, Red-crowned Pigeon)

219. **Ptilinopus coronulatus**
Coroneted Fruit-Dove

Ptilonopus coronulatus G. R. Gray, 1858, Proc. Zool. Soc. London: 185, pl. 138.

(Little Coroneted Fruit Dove, Diadem Fruit-dove)

220. **Ptilinopus pulchellus**
Beautiful Fruit-Dove

Columba pulchella Temminck, 1835, Pl. Col., livr. 95: 564.

221. **Ptilinopus rivoli**
White-breasted Fruit-Dove

Columba Rivoli Prévost, 1839-1843 (1843), in Knip, Les Pigeons, éd. 2, 2: 107, pl. 57.

(High Mountain Fruit Dove)

222. **Ptilinopus solomonensis**
Yellow-bibbed Fruit-Dove

Ptilonopus solomonensis G. R. Gray, 1870, Ann. Mag. Nat. Hist., ser. 4, 5: 328.

(Splendid Fruit Dove)

223. **Ptilinopus viridis**
Claret-breasted Fruit-Dove

Columba viridis Linnaeus, 1766, Syst. Nat., ed. 12, 1: 283.

(Red-breasted Fruit Dove)

224. **Ptilinopus iozonus**
Orange-bellied Fruit-Dove

Ptilonopus iozonus G. R. Gray, 1858, Proc. Zool. Soc. London: 186.

Note: There is a single unconfirmed report of *Ptilinopus insolitus* from the New Guinean region (Berggy 1978).

225. **Ptilinopus nanus** Dwarf Fruit-Dove

Columba naina Temminck, 1835, Pl. Col. livr., 95, pl. 565; emended to *nanus* by Temminck, 1838 (1839), Tabl. Méth.: 80.

(*Ptilinopus naina*, Little, Least, or Small Green Fruit Dove)

226. **Ducula concinna**
Elegant Imperial Pigeon

Carpophaga concinna Wallace, 1865, Ibis 1, new series: 383.

(Elegant Fruitpigeon, Blue-tailed Imperial Pigeon)

227. **Ducula pacifica** Pacific Imperial Pigeon

Columba pacifica Gmelin, 1789, Syst. Nat. 1: 777.

(Pacific Fruitpigeon)

228. **Ducula myristicivora**
Spice Imperial Pigeon

Columba (myristicivora) Scopoli, 1786, Delic. Flor. Faun. Insubr. 2: 94.

(Island Imperial Pigeon, Black-knobbed Fruit-Pigeon, or Spice Pigeon)

229. **Ducula rufigaster**
Purple-tailed Imperial Pigeon

Columba rufigaster Quoy and Gaimard, 1830, Voy. Astrolabe, Zool. 1: 245.

(Rufous-bellied Fruitpigeon, Purple-tailed Fruitpigeon)

230. **Ducula chalconota** Rufescent Imperial Pigeon

Carpophaga chalconota Salvadori, 1874, Ann. Mus. Civ. Stor. nat. Genova 6: 87.

(Red-breasted Imperial Pigeon, Rufous-breasted or Mountain Red-breasted Fruitpigeon)

231. **Ducula pistrinaria**
Grey Imperial Pigeon

Ducula pistrinaria Bonaparte 1855, Consp. Gen. Av. 2: 34.

(Island Imperial Pigeon, Floury Fruitpigeon)

232. **Ducula pinon** Pinon Imperial Pigeon

Columba Pinon Quoy and Gaimard, 1824, Voy. Uranie, Zool.: 118, Atlas: pl. 28.

(Black-shouldered Fruitpigeon)

233. **Ducula muellerii** Collared Imperial Pigeon

Columba mullerii Temminck, 1835, Pl. Col., livr. 96: 566.

(Muller's Imperial Pigeon, Black-collared Fruit Pigeon, Collared Fruitpigeon)

234. **Ducula zoeae** Zoe Imperial Pigeon

Columba Zoeae Lesson, 1826, Dict. Sci. Nat., éd. Levrault 40: 314.

(Black-belted Fruitpigeon, Banded Imperial Pigeon)

235. **Ducula bicolor** Pied Imperial Pigeon

Columba (bicolor) Scopoli, 1786, Del. Flor. Faun. Insubr. 2: 94.

This species now includes *D. spilorrhoa* as suggested by Johnstone (1981).

(*Myristicivora spilorrhoa*, *Carpophaga spilorrhoa*, Torres Strait Imperial Pigeon, Nutmeg Pigeon, Eastern, Western, or White Nutmeg Pigeon)

236. **Gymnophaps albertisii** Papuan Mountain Pigeon

Gymnophaps albertisii Salvadori, 1874, Ann. Mus. Civ. Stor. nat. Genova 6: 86.

(Mountain Pigeon, Bare-eyed Pigeon, D'Albertis' Pigeon)

ORDER **PSITTACIFORMES**

FAMILY **PSITTACIDAE:** PARROTS, LORIES AND ALLIES

Treatment follows the sequence, species and generic designations of Forshaw (1973), with emendations. In considering this assemblage to constitute only one family, we follow Smith (1975).

237. **Chalcopsitta atra** Black Lory

Psittacus (ater) Scopoli, 1786, Delic. Flor. Faun. Insubr. 2: 87.

(Rajah Lory)

238. **Chalcopsitta duivenbodei** Brown Lory

Chalcopsittacus Duivenbodei Dubois, 1884, Bull. Mus. Roy. d'Hist. nat. Belgique 3: 113, pl. 5.

(Duyvenbode's Lory)

239. **Chalcopsitta scintillata** Greater Streaked Lory

Psittacus sintillatus Temminck, 1835, Pl. Col., livr. 96, pl. 569; emended to *scintillatus* by Temminck, 1838 (1839), Table Méthod.: 61.

(Yellow-streaked Lory)

240. **Eos squamata** Moluccan Red Lory

Psittacus Squamatus Boddaert, 1783, Table Pl. Enlum.: 42.

(Violet-necked Lory or Moluccas Red Lory)

241. **Eos cyanogenia** Biak Red Lory

Eos cyanogenia Bonaparte, 1850, Compt. Rend. Acad. Sci., Paris 30: 135.

(Black-winged Lory)

242. **Pseudeos fuscata** Dusky Lory

Eos fuscata Blyth, 1858, Journ. Asiat. Soc. Bengal 27: 279.

(Dusk-orange Lory)

243. **Trichoglossus haematodus** Rainbow Lorikeet

Psittacus haematod[us] Linnaeus, 1771, Mantissa Plant.: 524.

(*Trichoglossus moluccanus*, *T. haematod*, Coconut or Rainbow Lory)

244. **Trichoglossus goldiei** Goldie's Lorikeet

Trichoglossus goldiei Sharpe, 1882, Journ. Linn. Soc. London, Zool. 16: 317, 426.

(*Psitteuteles goldiei*, Red-capped Streaked Lory)

245. **Lorius hypoinochrous**
Eastern Black-capped Lory

Lorius hypoinochrous G. R. Gray, 1859, List Bds. Brit. Mus., Psittacidae: 49.

(*Domicella hypoinochroa*, Purple-bellied Lory)

246. **Lorius lory** Western Black-capped Lory

Psittacus Lory Linnaeus, 1758, Syst. Nat., ed. 10, 1: 100.

(*Domicella lory*, Black-capped Lory)

247. **Charmosyna multistriata**
Streaked Lorikeet

Charmosynopsis multistriata Rothschild, 1911, Bull. Brit. Orn. Cl. 27: 45.

(Striated Lorikeet, Yellow Streaked Lory)

248. **Charmosyna wilhelminae**
Pygmy Lorikeet

Trichoglossus Wilhelminae A. B. Meyer, 1874, Journ. f. Ornith. 22: 55, 56.

(Pygmy Streaked Lory, Wilhelmina's Lorikeet)

249. **Charmosyna rubrigularis**
Red-chinned Lorikeet

Trichoglossus rubrigularis Sclater, 1881, Proc. Zool. Soc. London: 451.

(Red-chinned Lory)

250. **Charmosyna rubronotata** Red-fronted Lorikeet

Coriphilus rubronotatus Wallace, 1862, Proc. Zool. Soc. London: 165.

(Red-spotted Lorikeet, Red-fronted Blue-eared Lory)

251. **Charmosyna placentis**
Red-flanked Lorikeet

Psittacus placentis Temminck, 1834, Pl. Col., livr. 93: 553.

(Yellow-fronted Blue-eared Lory, Lowland Lorikeet)

252. **Charmosyna pulchella**
Little Red Lorikeet

Charmosyna pulchella G. R. Gray, 1859, List Bds. Brit. Mus., Psittacidae: 102.

(Fairy Lorikeet, Little Red Lory)

253. **Charmosyna josefinae**
Josephine's Lorikeet

Trichoglossus Josefinae Finsch, 1873, Atti Soc. Ital. Sci. Nat. Milano 15: 427, pl. 7.

(Josefine's Lory)

254. **Charmosyna papou** Papuan Lorikeet

Psittacus (Papou) Scopoli, 1786, Delic. Flor. Faun. Insubr. 2: 86.

(Fairy Lory, Stella's Lorikeet)

255. **Oreopsittacus arfaki**
Plum-faced Lorikeet

Trichoglossus (Charmosyna) Arfaki A. B. Meyer, 1874, Verh. Zool.-Bot. Ges. Wien 24: 37.

(Whiskered Lorikeet, Plum-faced Mountain Lory)

256. **Neopsittacus musschenbroekii**
Yellow-billed Lorikeet

Nanodes Musschenbroekii 'von Rosenberg' Schlegel, 1871 (1873), Nederl. Tijdschr. Dierk. 4: 34.

(Yellow-billed Mountain Lory, Musschenbroek's Lorikeet)

257. **Neopsittacus pullicauda** Orange-billed Lorikeet

Neopsittacus pullicauda Hartert, 1896, Novit. Zool. 3: 17.

(Orange-billed Mountain Lory, Emerald Lorikeet)

258. **Probosciger aterrimus** Palm Cockatoo

Psittacus aterrimus Gmelin, 1788, Syst. Nat. 1: 330.

259. **Cacatua galerita**
Sulphur-crested Cockatoo

Psittacus galeritus Latham, 1790, Index Ornith. 1: 109.

(*Kakatoe galerita*, White Cockatoo)

260. **Cacatua pastinator** Little Corella

Licmetis pastinator Gould, 1840 (1841), Proc. Zool. Soc. London: 175.

This treatment follows Schodde *et al.* (1979).

(*Cacatua sanguinea*)

261. **Micropsitta pusio** Buff-faced Pygmy-Parrot

Nasiterna pusio Sclater, 1865 (1866), Proc. Zool. Soc. London: 620, pl. 35.

(Buffy-faced Pygmy Parrot, Buff-faced Pygmyparrot)

262. **Micropsitta keiensis** Yellow-capped Pygmy-Parrot

Nasiterna keiensis Salvadori, 1875, Ann. Mus. Civ. Stor. nat. Genova 7: 984.

(Yellow-capped Pygmyparrot)

263. **Micropsitta geelvinkiana** Geelvink Pygmy-Parrot

Nasiterna pygmaea Geelvinkiana Schlegel, 1871 (1873), Nederl. Tijdschr. Dierk. 4: 7.

(Geelvink Pygmyparrot)

264. **Micropsitta bruijnii** Red-breasted Pygmy-Parrot

Nasiterna bruijnii Salvadori, 1875, Ann. Mus. Civ. Stor. nat. Genova 7: 715.

(Mountain Pygmy Parrot, Red-breasted Pygmyparrot)

265. **Cyclopsitta gulielmiterti** Orange-breasted Fig-Parrot

Psittacula gulielmi III Schlegel, 1866, Nederl. Tijdschr. Dierk. 3: 252.

We follow Storr (1973) and Schodde (1978) in reinstating *Cyclopsitta*. See also Holyoak (1970) for another view.

(*Opopsitta gulielmiterti, Psittaculirostris gulielmiterti*, William's or Black-cheeked Fig Parrot)

266. **Cyclopsitta diophthalma** Double-eyed Fig-Parrot

Psittacula diophthalmus Hombron and Jacquinot, 1841, Ann. Sci. Nat., Zool., ser. 2, 16: 318.

(*Psittaculirostris diophthalma, Opopsitta diophthalma*, Two-eyed Fig Parrot)

267. **Psittaculirostris desmarestii** Large Fig-Parrot

Psittacus Desmarestii Desmarest, 1826, Dict. Sci. Nat., éd. Levrault 39: 89.

(Desmarest's Figparrot)

268. **Psittaculirostris edwardsii** Edwards' Fig-Parrot

Cyclopsittacus Edwardsii Oustalet, 1885, Ann. Sci. Nat., Zool. ser. 6, 19 (art. 3): 1.

(Edward's Fig Parrot)

269. **Psittaculirostris salvadorii** Salvadori's Fig-Parrot

Cyclopsittacus Salvadorii Oustalet, 1880, Bull. Assoc. Sci. France, ser. 2, 1: 172.

270. **Psittacella brehmii** Brehm's Tiger-Parrot

Psittacella brehmii Schlegel, 1871 (1873) Nederl. Tijdschr. Dierk. 4: 35.

(Brehm's Parrot)

271. **Psittacella picta** Painted Tiger-Parrot

Psittacella picta Rothschild, 1896, Bull. Brit. Ornith. Cl. 6: 5.

(Painted Parrot, Timber-line Parrot)

272. **Psittacella modesta** Modest Tiger-Parrot

Psitacella [sic] *modesta* Schlegel, 1871 (1873), Nederl. Tijdschr. Dierk. 4: 36.

(Modest Parrot, Barred Little Parrot)

273. **Psittacella madaraszi** Madarasz's Tiger-Parrot

Psittacella madaraszi A. B. Meyer, 1886, Zeitschr. Ges. Ornith. 3: 4, pl. 1, fig. 1.

(Madarasz's Parrot, Plain-breasted Little Parrot)

274. **Geoffroyus geoffroyi**
Red-cheeked Parrot

Psittacus geoffroyi Bechstein, 1811, in Latham's Kurze Uebers. Vög. 4: 103, pl. 21.

275. **Geoffroyus simplex**
Blue-collared Parrot

Pionias simplex A. B. Meyer, 1874, Verh. Zool.-Bot. Ges. Wien 24: 39.

(Lilac-collared Parrot)

276. **Tanygnathus megalorynchos**
Great-billed Parrot

Psittacus megalorynchos Boddaert, 1783, Table Pl. Enlum.: 45.

(*T. megalorhynchus*, Moluccan or Island Parrot)

277. **Eclectus roratus** Eclectus Parrot

Psittacus roratus P. L. S. Müller, 1776, Natursyst., Suppl.: 77. See Mayr (1968a) for suppression of *Larius*.

(*Larius roratus*, Red-sided Eclectus Parrot, Red-sided Parrot, Kalanga)

278. **Psittrichas fulgidus** Vulturine Parrot

Banksianus fulgidus Lesson, 1830, Traité Ornith. 1: 181.

(Pesquet's Parrot)

279. **Alisterus chloropterus**
Papuan King-Parrot

Aprosmictus chloropterus Ramsay, 1879, Proc. Linn. Soc. New South Wales 3: 251.

(Green-winged King-parrot)

This and the following species may be conspecific (Diamond, *in litt.*). More study is needed of the distribution of these two forms in western New Guinea).

280. **Alisterus amboinensis**
Moluccan King-Parrot

Psittacus amboinensis Linnaeus, 1766, Syst. Nat., ed. 12, 1: 141.

(Amboina Kingparrot)

281. **Aprosmictus erythropterus**
Red-winged Parrot

Psittacus erythropterus Gmelin, 1788, Syst. Nat. 1: 343.

282. **Loriculus aurantiifrons**
Papuan Hanging Parrot

Loriculus aurantiifrons Schlegel, 1871 (1873) Nederl. Tijdschr. Dierk. 4: 9.

(Bat Lorikeet, Orange-fronted Hangingparrot)

ORDER **CUCULIFORMES**

FAMILY **CUCULIDAE:** CUCKOOS, KOELS AND COUCALS

General treatment follows Peters (1940) with emendations.

283. **Cuculus saturatus** Oriental Cuckoo

Cuculus saturatus 'Hodgson', Blyth, 1843, Journ. Asiat. Soc. Bengal, 12 (2): 942.

(Blyth's or Himalayan Cuckoo)

Note: Earlier lists included *C. canorus* on the basis of misidentified specimens of *saturatus* (Mees 1979).

284. **Cuculus pallidus** Pallid Cuckoo

Columba pallida Latham, 1801, Index Ornith., Suppl.: 60.

See Stronoch (1981b) and Mees (1982b) for additional records from the region.

285. **Cacomantis variolosus** Brush Cuckoo

Cuculus variolosus Vigors and Horsfield, 1826 (1827), Trans. Linn. Soc. London 15(1): 300.

(*Cuculus variolosus*, Grey-breasted Brush Cuckoo)

286. **Cacomantis castaneiventris** Chestnut-breasted Cuckoo

Cuculus (Cacomantis) castaneiventris Gould, 1867, Ann. Mag. Nat. Hist., ser. 3, 20: 269.

(*Cuculus castaneiventris*, Chestnut-breasted Brush Cuckoo)

287. **Cacomantis flabelliformis** Fan-tailed Cuckoo

Cacomantis flabelliformis Latham, 1801, Index Ornith., Suppl.: 30.

Specific nomenclature follows Mason (1982), who shows that *flabelliformis* is the oldest valid name for this taxon.

(*Cacomantis pyrrhophanus, Cuculus pyrrhophanus*, Fan-tailed Brush Cuckoo)

288. **Rhamphomantis megarhynchus** Long-billed Cuckoo

Cuculus megarhynchus G. R. Gray, 1858, Proc. Zool. Soc. London: 184.

(Little Long-billed Cuckoo)

289. **Chrysococcyx osculans** Black-eared Cuckoo

Chalcites osculans Gould, 1847, Proc. Zool. Soc. London: 32.

Generic assignment follows Condon (1975).

(*Misocalius osculans*)

290. **Chrysococcyx basalis** Horsfield's Bronze-Cuckoo

Cuculus basalis Horsfield, 1821, Trans. Linn. Soc. London 13(1): 179.

Generic placement of this and the following congeners follows Friedmann (1968).

(*Chalcites basalis*, Narrow-billed or Rufous-tailed Bronze Cuckoo)

291. **Chrysococcyx lucidus** Shining Bronze-Cuckoo

Cuculus lucidus Gmelin, 1788, Syst. Nat. 1 (1): 421.

The name Golden Bronze-Cuckoo is available for the race *plagosus*, the most common population to visit New Guinea.

(*Chalcites lucidus*)

292. **Chrysococcyx minutillus** Malay Bronze-Cuckoo

Chrysococcyx minutillus Gould, 1859, Proc. Zool. Soc. London: 128.

Parker (1981) has determined that the name *malayanus* Raffles (1822) is a junior synonym of *C. xanthorhynchus*, and that the next available name for the species is *minutillus*. We include the populations *russatus*, *minutillus*, and *poecilurus* as races of this species, but further study is needed (Ford 1981c, Parker 1981). We use the name Malay Bronze-Cuckoo for this polytypic species; the names Little and Gould's bronze-cuckoos are reserved for the distinctive races *poecilurus* and *russatus* respectively.

(*Chalcites malayanus*, Little or Gould's Bronze Cuckoo, Rufous Bronze-Cuckoo).

293. **Chrysococcyx ruficollis** Rufous-throated Bronze-Cuckoo

Lamprococcyx ruficollis Salvadori, 1875, Ann. Mus. Civ. Stor. nat. Genova 7: 913.

(*Chalcites ruficollis*, Reddish-throated or Mountain Bronze Cuckoo)

294. **Chrysococcyx meyerii** White-eared Bronze-Cuckoo

Chrysococcyx meyerii Salvadori, 1874, Ann. Mus. Civ. Stor. nat. Genova 6: 82.

(*Chalcites meyerii*, Myer's [sic] Bronze Cuckoo, Meyer's Bronze Cuckoo, White-eared Cuckoo)

295. **Caliechthrus leucolophus** White-crowned Koel

Cuculus leucolophus S. Müller, 1840 (1839-1844), Verh. Nat. Gesch. Nederl., Land-en Volkenkunde: 22.

296. **Microdynamis parva** Dwarf Koel

Eudynamis parva Salvadori, 1875, Ann. Mus. Civ. Stor. nat. Genova 7: 986.

Mayr (1944) suggested including *Microdynamis* in *Eudynamys*.

(*Eydynamys parva, Eudynamis parva*, Black-capped Cuckoo)

297. **Eudynamys scolopacea** Common Koel
Cuculus scolopaceus Linnaeus, 1758, Syst. Nat. ed. 10, 1: 111.
There is some question about the specific status of a number of populations of this group inhabiting Wallacea, New Guinea, and Australia (Rand 1941, M. D. Bruce, *in litt.*). It has been suggested by Rand that the strikingly different female plumages of the forms *rufiventer* and *subcyanocephala* indicate that these populations have achieved specific status. More study is needed.
(*Eudynamis scolopacea*, Koel, Indian Koel)

298. **Eudynamys taitensis** Long-tailed Koel
Cuculus taitensis Sparrman, 1787, Mus. Carlson. 2: pl. 32.
(*Eudynamis taitensis, Urodynamis taitensis*)

299. **Scythrops novaehollandiae** Channel-billed Cuckoo
Scythrops novaehollandiae Latham, 1790, Index Ornith. 1: 141.

300. **Centropus menbeki** Greater Black Coucal
Centropus menbeki Lesson 1828 (1829), Voy. Coquille, Atlas, pl. 33; Zool. 1 (13): 600.
(Menbek's or Greater Coucal, Black Jungle Coucal)

301. **Centropus chalybeus** Biak Coucal
Nesocentor chalybeus Salvadori, 1875, Ann. Mus. Civ. Stor. nat. Genova 7: 915.
(Biak Island Coucal, Biak Bronze Coucal)

302. **Centropus phasianinus** Pheasant Coucal
Cuculus phasianinus Latham, 1801, Index Ornith., Suppl.: 30.
(*Centropus phasianus* [sic])

303. **Centropus bernsteini** Lesser Black Coucal
Centropus bernsteini Schlegel, 1866, Nederl. Tijdschr. Dierk. 3: 251.
(*Centropus bernsteinii*, Black Scrub Coucal, Bernstein's Coucal)

ORDER **STRIGIFORMES**

FAMILY **TYTONIDAE:** BARN OWLS

Treatment follows Schodde & Mason (1980).

304. **Tyto tenebricosa** Sooty Owl
Strix tenebricosus Gould, 1845, Proc. Zool. Soc. London: 80.

305. **Tyto novaehollandiae** Masked Owl
Strix (?) *Novae Hollandiae* Stephens, 1826, in Shaw's Gen. Zool. 13 (2): 61.

306. **Tyto alba** Barn Owl
Tyto alba Scopoli, 1769, Annus 1, Hist.-Nat.: 21.

307. **Tyto capensis** Grass Owl
Strix capensis A. Smith 1834, S. Afr. Quart. Journ. (2): 317.
Condon (1975) considers that the Australasian populations constitute a distinct species. *Tyto longimembris*. We follow Amadon (1959), Storr (1973), Parker (1977) and Schodde & Mason (1980) in placing these populations with *capensis*.

FAMILY **STRIGIDAE:** TYPICAL OWLS

Treatment follows Peters (1940) with emendations, as noted.

308. **Otus magicus** Moluccan Scops-Owl
Strix magica S. Müller, 1841, Verh. Nat. Gesch. Nederl., Land-en Volkenkunde: 110.
Treatment follows Marshall (1978). Previous authorities gave this insular population specific status.
(*Otus beccarii*, Biak or Papuan Scops Owl)

309. **Uroglaux dimorpha** Papuan Hawk-Owl
Athene dimorpha Salvadori, 1874, Ann. Mus. Civ. Stora nat. Genova 6: 308.

310. **Ninox rufa** Rufous Owl

Athene rufa Gould, 1846, Proc. Zool. Soc. London: 18.

(Barred Papuan Owl)

311. **Ninox connivens** Barking Owl

Falco connivens Latham, 1801, Index Ornith., Suppl: 12.

(Winking Owl, Barking Hawk-Owl)

312. **Ninox novaeseelandiae** Southern Boobook

Strix novae Seelandiae Gmelin, 1788, Syst. Nat. 1: 296.

We follow Mees (1982a) in considering this bird as *N. novaeseelandiae*. For a differing opinion on the nomenclature of this species, see Schodde & Mason (1980).

(*Ninox boobook*, Boobook Owl, Common Boobook, Spotted Owl)

313. **Ninox theomacha** Papuan Boobook

Spiloglaux theomacha Bonaparte, 1855, Compt. Rend. Acad. Sci., Paris 41: 654.

(Jungle Boobook, Brown Owl)

ORDER **CAPRIMULGIFORMES**

FAMILY **PODARGIDAE:** FROGMOUTHS

Treatment follows Schodde & Mason (1980).

314. **Podargus ocellatus**
Marbled Frogmouth

Podargus ocellatus Quoy and Gaimard, 1830, Voy. Astrolabe, Zool. 1: 208; Atlas, Ois.: pl. 14.

(Little Papuan Frogmouth)

315. **Podargus papuensis**
Papuan Frogmouth

Podargus papuensis Quoy and Gaimard, 1830, Voy. Astrolabe, Zool. 1: 207; Atlas, Ois.: pl. 13.

(Great Papuan Frogmouth)

Note: There is a single unconfirmed sight record of *P. strigoides* from near Morehead, Western Province, Papua New Guinea (Mackay 1976).

FAMILY **AEGOTHELIDAE:** OWLET-NIGHTJARS

Treatment follows Peters (1940) with emendations as indicated.

316. **Aegotheles insignis**
Feline Owlet-nightjar

Aegotheles insignis Salvadori, 1875, Ann. Mus. Civ. Stor. nat. Genova 7: 916.

(*Euaegotheles insignis*, Large Owlet-nightjar)

317. **Aegotheles cristatus**
Australian Owlet-nightjar

Caprimulgus cristatus J. White, 1790, J. Voy. New South Wales: 241 and pl.

(Owlet-nightjar, Savannah Owlet-nightjar)

318. **Aegotheles bennettii**
Barred Owlet-nightjar

Aegotheles bennettii Salvadori and D'Albertis, 1875, Ann. Mus. Civ. Stor. nat. Genova 7: 816.

(Collared Owlet-nightjar)

319. **Aegotheles wallacii**
Wallace's Owlet-nightjar

Aegotheles wallacii G. R. Gray, 1859, Proc. Zool. Soc. London: 154.

(White-spotted Owlet-nightjar)

320. **Aegotheles albertisi**
Mountain Owlet-nightjar

Aegotheles albertisi Sclater, 1873 (1874), Proc. Zool. Soc. London: 696.

(*A. albertisii*)

321. **Aegotheles archboldi**
Archbold's Owlet-nightjar

Aegotheles albertisii archboldi Rand, 1941, Am. Mus. Novit. 1102: 10.

Shown to be a full species by Junge (1953). Previous lists treated this form as a race of *albertisi*.

FAMILY **CAPRIMULGIDAE:** NIGHTJARS

Treatment follows Peters (1940) and Schodde & Mason (1980) with emendations.

322. **Eurostopodus argus** Spotted Nightjar

Caprimulgus argus Hartert, 1892, Cat. Bds. Brit. Mus. 14: 308.

Nomenclature follows Schodde & Mason (1980).

(*Eurostopodus guttatus, Caprimulgus guttatus*)

323. **Eurostopodus mystacalis** White-throated Nightjar

Caprimulgus mystacalis Temminck, 1826, Pl. Col., livr. 69: pl. 410.

Treatment follows Mayr (1941) and Condon (1975).

(*Eurostopodus albogularis, Caprimulgus mystacalis*)

324. **Eurostopodus papuensis** Papuan Nightjar

Caprimulgus papuensis Schlegel, 1866, Nederl. Tijdschr. Dierk. 3: 340.

325. **Eurostopodus archboldi** Mountain Nightjar

Lyncornis archboldi Mayr and Rand, 1935, Am. Mus. Novit. 814: 4.

(Archbold's Nightjar)

326. **Caprimulgus indicus** Jungle Nightjar

Caprimulgus indicus Latham, 1790, Index Ornith. 2: 588.

(Grey or Japanese Nightjar, Indian Jungle Nightjar)

327. **Caprimulgus macrurus** Large-tailed Nightjar

Caprimulgus macrurus Horsfield, 1821, Trans. Linn. Soc. London 13 (1): 142.

(White-tailed or Long-tailed Nightjar, Coffin Bird)

ORDER **APODIFORMES**

Treatment generally follows Peters (1940), with emendations. Family sequence places Hemiprocnidae before Apodidae, following Zusi & Bentz (1982).

FAMILY **HEMIPROCNIDAE:** TREE-SWIFTS

328. **Hemiprocne mystacea** Moustached Tree-swift

Cypselus mystaceus Lesson, 1827 (1830), Voy. Coquille, Atlas: pl. 22.

(Moustached Swift)

FAMILY **APODIDAE:** SWIFTLETS AND SWIFTS

Treatment follows Condon (1975), and Salomonsen (1983) for the Melanesian forms.

329. **Collocalia vanikorensis** Uniform Swiftlet

Hirundo vanikorensis Quoy and Gaimard, 1830, Voy. Astrolabe, Zool. 1: 206; Atlas Ois. pl. 12, fig. 3.

This and the following three species are placed in *Aerodramus* by Medway & Pye (1977).

(*Aerodramus vanikorensis*, Lowland Swiftlet)

330. **Collocalia hirundinacea** Mountain Swiftlet

Collocalia fuciphaga hirundinacea Stresemann, 1914, Verh. Ornith. Ges. Bayern 12: 7.

(*Aerodramus hirundinacea*)

331. **Collocalia whiteheadi** Whitehead's Swiftlet

Collocalia whiteheadi Ogilvie-Grant, 1894, Ibis: 459.

For treatment of species limits of the large Papuan *Collocalia* see Somadikarta (1967) and Ripley (1983). Here, all of the large four-toed *Collocalia* are considered conspecific with *whiteheadi*, while all three-toed populations are considered to constitute *papuensis*.

(*Collocalia brevirostris, Aerodramus brevirostris, C. nuditarsus*)

332. **Collocalia papuensis** Three-toed Swiftlet

Collocalia whiteheadi papuensis Rand, 1941, Am. Mus. Novit. 1102: 10.

Originally this population was included in *whiteheadi*, a four-toed form (Somadikarta 1967).

333. **Collocalia esculenta** Glossy Swiftlet

Hirundo esculenta Linnaeus, 1758, Syst. Nat., ed. 10, 1: 191.

(White-bellied Swiftlet)

334. **Mearnsia novaeguineae** Papuan Spine-tailed Swift

Chaetura novae guineae D'Albertis and Salvadori, 1879, Ann. Mus. Civ. Stor. nat. Genova 14: 55.

(Chaetura novaeguinae, New Guinea Spine-tailed Swift)

335. **Hirundapus caudacutus** White-throated Needletail

Hirundo caudacuta Latham, 1801, Index Ornith., Suppl.: 57.

See Collins & Brooke (1976) for notes on generic treatment.

(*Chaetura caudacuta*, Spine-tailed Swift, Northern Spine-tailed Swift)

336. **Apus pacificus** Fork-tailed Swift

H[irundo] pacifica Latham, 1801, Index Ornith., Suppl.: 58.

(White-rumped Swift, Pacific Swift, Large Whiterumped Swift)

ORDER **CORACIIFORMES:**

FAMILY **ALCEDINIDAE** KINGFISHERS AND KOOKABURRAS

Treatment follows Fry (1980).

337. **Tanysiptera galatea** Common Paradise-Kingfisher

Tanysiptera galatea G. R. Gray, 1859, Proc. Zool. Soc. London: 154.

Fry (1980) included the populations from Biak I. (*riedelii*) and Kofiau I. (*ellioti*) in this species. The two island forms have previously been considered full species, members of the *galatea* superspecies.

(Biak Paradise-Kingfisher, Kofiau Paradise-Kingfisher)

338. **Tanysiptera carolinae** Numfor Paradise-Kingfisher

Tanysiptera carolinae Schlegel, 1871 (1873), Nederl. Tijdschr. Dierk. 4: 13.

339. **Tanysiptera hydrocharis** Little Paradise-Kingfisher

Tanysiptera hydrocharis G. R. Gray, 1858, Proc. Zool. Soc. London: 172.

(Aru Paradise Kingfisher)

340. **Tanysiptera sylvia** Australian Paradise-Kingfisher

Tanysiptera sylvia Gould, 1850, in Jardine's Contr. Ornith.: 105.

(White-tailed Kingfisher, Buff-breasted or White-tailed Paradise Kingfisher)

341. **Tanysiptera nympha** Red-breasted Paradise-Kingfisher

Tanysiptera nympha G. R. Gray, 1841, Ann. and Mag. Nat. Hist., ser. 1, 6: 237.

(Pink-breasted Paradise Kingfisher)

342. **Tanysiptera danae** Brown-headed Paradise-Kingfisher

Tanysiptera danae Sharpe, 1880, Ann. Mag. Nat. Hist., ser. 5, 6: 231.

(Brown-backed Paradise Kingfisher)

343. **Melidora macrorrhina** Hook-billed Kingfisher

Dacelo macrorrhinus Lesson, 1827, Bull. Sci. Nat. Férussac 12: 131.

(*Melidora macrorhina*, Hook-billed Kookaburra)

344. **Clytoceyx rex** Shovel-billed Kingsfisher
Clytoceyx rex Sharpe, 1880, Ann. Mag. Nat. Hist., ser. 5, 6: 231.
(Shovel-billed Kookaburra, Crab-eating Kingfisher)

345. **Dacelo gaudichaud**
Rufous-bellied Kookaburra
Dacelo Gaudichaud Quoy and Gaimard, 1824, Voy. Uranie, Zool. 1: 112, pl. 25.
(*Sauromarptis gaudichaud, Dacelo gaudichaudi*, Rufous-bellied Giant Kingfisher, Gaudichaud's Kingfisher)

346. **Dacelo tyro** Spangled Kookaburra
Dacelo tyro G. R. Gray, 1858, Proc. Zool. Soc. London: 171, pl. 133.
(*Sauromarptis tyro*, Aru Giant Kingfisher)

347. **Dacelo leachii** Blue-winged Kookaburra
[Dacelo] leachii Vigors and Horsfield, 1826 (1827), Trans. Linn. Soc. London 15 (1): 205.

348. **Halcyon nigrocyanea**
Blue-black Kingfisher
Halcyon nigrocyanea Wallace, 1862, Proc. Zool. Soc. London: 165, pl. 19.
(Black-sided Kingfisher)

349. **Halcyon macleayii** Forest Kingfisher
Halycon macleayii Jardine and Selby, 1830, Illustr. Ornith. 2 (7): pl. 101.

350. **Halcyon chloris** Collared Kingfisher
Alcedo chloris Boddaert, 1783, Table Pl. Enlum.: 49.
(Mangrove, White collared or Common Kingfisher)

351. **Halcyon sancta** Sacred Kingfisher
Halcyon [sancta] Vigors and Horsfield, 1827, Trans. Linn. Soc. London 15 (1): 206.

352. **Halcyon saurophaga** Beach Kingfisher
Halcyon saurophaga Gould, 1843, Proc. Zool. Soc. London: 103.
(White-headed Kingfisher)

353. **Halcyon torotoro**
Yellow-billed Kingfisher
Syma torotoro Lesson, 1827, Bull. Sci. Nat. Férussac 11: 443.
(Lesser or Lowland Yellow-billed Kingfisher)

354. **Halcyon megarhyncha**
Mountain Kingfisher
Syma megarhyncha Salvadori, 1896, Ann. Mus. Civ. Stor. nat. Genova, ser. 2, 16[=36]: 70.
(Greater or Mountain Yellow-billed Kingfisher)

355. **Ceyx lepidus** Dwarf Kingfisher
Ceyx lepida Temminck, 1836, Pl. Col., livr. 100: pl. 595, fig. 1.
(*Alcedo lepida*, Variable Dwarf Kingfisher)

356. **Alcedo pusilla** Little Kingfisher
Ceyx pusilla Temminck, 1836, Pl. Col., livr.: 100, pl. 595, fig. 3.
(*Alcyone pusilla, Ceyx pusillus*)

357. **Alcedo azurea** Azure Kingfisher
Alcedo azurea Latham, 1801, Index Ornith., Suppl.: 32.
(*Alcyone azurea, Ceyx azureus*)

358. **Alcedo atthis** Common Kingfisher
Gracula Atthis Linnaeus, 1758 Syst. Nat., ed. 10, 1: 109.
(River or Eurasian Kingfisher, Kingfisher)

FAMILY **MEROPIDAE:** BEE-EATERS

Treatment follows Fry (1969).

359. **Merops philippinus**
Blue-tailed Bee-eater
Merops philippinus Linnaeus, 1766, Syst. Nat., ed. 12, 1: errata at end of the volume; name for *Merops* fifth species, p. 183.

Some authorities treat this form as being conspecific with *Merops superciliosus*, but we agree with Mees (1982b) who cites the study of Marien (1950) as evidence for maintaining full specific status for *philippinus*.

360. **Merops ornatus** Rainbow Bee-eater
Merops ornatus Latham, 1801, Index Ornith., Suppl.: 35.
(Rainbow Bird)

FAMILY **CORACIIDAE:** ROLLERS

Treatment follows Peters (1945).

361. **Eurystomus orientalis** Dollarbird
Eurystomus orientalis Linnaeus, 1766, Syst. Nat. ed. 12, 1: 159.
(Broad-billed Roller)

FAMILY **BUCEROTIDAE:** HORNBILLS

Treatment follows Sanft (1960). See also Kemp (1978).

362. **Rhyticeros plicatus** Blyth's Hornbill
Buceros plicatus J. R. Forster, 1781, Indische Zool.: 40.
(*Aceros plicatus*, Kokomo, Papuan Hornbill)

ORDER **PASSERIFORMES**

There is little agreement over the best familial sequence of perching birds—especially for the poorly known Australasian groups. We generally follow the Peters sequence with a variety of emendations, noted where they are made. Sibley & Ahlquist (1982a, b, c, in press) and Sibley *et al.* (1984) have offered a radically revised outlook on the familial organization of the songbirds, based on the biochemical technique of DNA:DNA hybridization. We follow several of their suggested familial changes.

FAMILY **PITTIDAE:** PITTAS

Treatment follows Mayr (1979a).

363. **Pitta erythrogaster** Blue-breasted Pitta
Pitta erythrogaster Temminck, 1823, Pl. Col., livr. 36, pl. 212.
(Red-bellied Pitta)

364. **Pitta sordida** Hooded Pitta
Turdus sordidus P. L. S. Müller, 1776, Natursyst., Suppl.: 143.
(Black-headed Pitta)

365. **Pitta versicolor** Noisy Pitta
Pitta versicolor Swainson, 1825, Zool. Journ. 1: 468.
(Buff-breasted Pitta)

FAMILY **ALAUDIDAE:** LARKS

Treatment follows Peters (1960).

366. **Mirafra javanica** Singing Bushlark
Mirafra javanica Horsfield, 1821, Trans. Linn. Soc. London 13 (1): 159.
(Bushlark)

FAMILY **HIRUNDINIDAE:** SWALLOWS AND MARTINS

Treatment generally follows Peters (1960) with emendations by Phillips (1973) and Schodde (1975).

367. **Riparia riparia** Sand Martin
Hirundo riparia Linnaeus, 1758, Syst. Nat., ed. 10, 1: 192.
There are two sight records for Papua New Guinea (Thompson 1979, Bell 1979).
(Bank Swallow, Collared Sand Martin)

368. **Hirundo rustica** Barn Swallow
Hirundo rustica Linnaeus, 1758, Syst. Nat., ed. 10, 1: 191.
(Swallow, Common or European Swallow)

369. **Hirundo tahitica** Pacific Swallow
Hirundo tahitica Gmelin, 1789, Syst. Nat. 1: 1016.
(House Swallow)

370. **Hirundo neoxena** Welcome Swallow

Hirundo neoxena Gould, 1842 (1843), Proc. Zool. Soc. London: 131.

First recorded in New Guinea by Finch (1980c).

371. **Hirundo daurica** Red-rumped Swallow

Hirundo daurica Linnaeus, 1771, Mantissa Plant.: 528.

Specific treatment follows King & Dickinson (1975). Some authorities consider the population *striolata* a full species. First recorded in the region by Filewood (1974).

(*Cecropis daurica, Hirundo striolata,* Striated Swallow)

372. **Hirundo nigricans** Tree Martin

Hirundo nigricans Vieillot, 1817, Nouv. Dict. Hist. Nat., nouv. éd. 14: 523.

(*Cecropis nigricans, Petrochelidon nigricans, Hylochelidon nigricans*)

373. **Hirundo ariel** Fairy Martin

Collocalia Ariel Gould, 1842 (1843), Proc. Zool. Soc. London: 132.

First recorded in the region by Bell (1968).

(*Petrochelidon ariel, Cecropis ariel*)

FAMILY **MOTACILLIDAE:** WAGTAILS AND PIPITS

Treatment follows Vaurie *et al.* (1960).

374. **Motacilla flava** Yellow Wagtail

Motacilla flava Linnaeus, 1758, Syst. Nat. ed. 10, 1: 185.

375. **Motacilla cinerea** Grey Wagtail

Motacilla cinerea Tunstall, 1771, Orn. Brit.: 2.

(*Motacilla caspica*)

376. **Anthus novaeseelandiae** Richard's Pipit

Alauda novae Seelandiae Gmelin, 1789, Syst. Nat. 1 (2): 799.

(*Anthus australis,* Australian Pipit)

377. **Anthus gutturalis** Alpine Pipit

Anthus gutturalis De Vis, 1894, Ann. Rept. Brit. New Guinea, 1893-1894: 103.

(New Guinea Pipit)

FAMILY **CAMPEPHAGIDAE:** CUCKOO-SHRIKES AND TRILLERS

Treatment follows Peters *et al.* (1960) with minor emendation. The sequence here at the first three species in *Coracina* follows Mayr (1941).

378. **Coracina longicauda** Hooded Cuckoo-shrike

Graucalus longicauda De Vis, 1890, Ann. Rept. Brit. New Guinea, 1888-1889: 59.

(Black-hooded Greybird)

379. **Coracina papuensis** White-bellied Cuckoo-shrike

Corvus papuensis Gmelin, 1788, Syst. Nat. 1: 371.

(Papuan Greybird, Little Cuckoo-shrike)

380. **Coracina novaehollandiae** Black-faced Cuckoo-shrike

Turdus novae Hollandiae Gmelin, 1789, Syst. Nat. 1: 814.

(Australian Greybird, Large Cuckoo-shrike)

381. **Coracina caeruleogrisea** Stout-billed Cuckoo-shrike

Campephaga caeruleogrisea G. R. Gray, 1858, Proc. Zool. Soc. London: 179.

(Stout-billed Greybird)

382. **Coracina lineata** Yellow-eyed Cuckoo-shrike

Ceblepyris lineata Swainson, 1825, Zool. Journ. 1: 466.

(Barred or Yellow-eyed Greybird, Barred Cuckoo-shrike)

383. **Coracina boyeri** Boyer's Cuckoo-shrike

Campephaga Boyeri G. R. Gray, 1846, Gen. Bds. 1: [283].

(Rufous-underwing Greybird)

384. **Coracina tenuirostris** Cicadabird

Gracaulus [sic] *tenuirostris* Jardine, 1831, Edinburgh Journ. Nat. Geog. Sci., new series, 3: 211.

(*Edolisoma tenuirostre*, Slender-billed or Long-billed Cuckooshrike, Long-billed Greybird, Jardine Triller)

385. **Coracina morio**
Black-shouldered Cuckoo-shrike

Ceblepyris morio S. Müller, 1843, Verh. Nat. Gesch., Land-en Volkenkunde: 189.

White & Bruce (1985) treat the population *incerta* as a full species distinct from *morio.*

(*Edolisoma morio*, Muller's Cuckoo-shrike or Greybird)

386. **Coracina schisticeps**
Grey-headed Cuckoo-shrike

Campephaga schisticeps G. R. Gray, 1846, Gen. Bds. 1: 283 (ex Hombron and Jacquinot, Vôy. Pole Sud).

(*Edolisoma schisticeps*, Slaty Cuckooshrike, Gray's Greybird)

387. **Coracina melaena** Black Cuckoo-shrike

Lanius melas Lesson, 1828, Man. Ornith. 1: 128.

(*Edolisoma melan*, Black Greybird)

388. **Coracina montana**
Black-bellied Cuckoo-shrike

Campephaga montona [sic] Meyer, 1874, Sitzungb. Akad. Wiss. Wien 69: 386.

(*Edolisoma montanum*, Black-bellied Greybird)

389. **Campochaera sloetii**
Golden Cuckoo-shrike

Campephaga sloetii Schlegel, 1866, Nederl. Tijdsch. Dierk. 3: 253.

(Golden Triller, Orange Cuckoo-shrike)

390. **Lalage sueurii** White-winged Triller

Turdus Suerii [sic] Vieillot, 1818, Nouv. Dict. Hist. Nat., nouv. éd. 20: 270.

(Australian Triller)

391. **Lalage atrovirens** Black-browed Triller

Campephaga (Lalage) atrovirens G. R. Gray, 1861 (1862), Proc. Zool. Soc. London: 430.

392. **Lalage leucomela** Varied Triller

[Campephaga] Leucomela Vigors and Horsfield, 1827, Trans. Linn. Soc. London 15: 215.

(White-browed or Pied Triller)

FAMILY **LANIIDAE:** SHRIKES

Treatment follows Rand (1960).

393. **Lanius cristatus** Brown Shrike

Lanius cristatus Linnaeus, 1758, Syst. Nat., ed. 10, 1: 93.

Included in the addendum of Rand & Gilliard (1967) based on a single specimen record of *L. cristatus lucionensis.*

394. **Lanius schach** Long-tailed Shrike

Lanius schach Linnaeus, 1758, Syst. Nat., ed. 10, 1: 94.

(Red-backed, Rufousbacked, Black-headed, Black-capped or Schach Shrike)

FAMILY **TURDIDAE:** THRUSHES

Treatment follows Ripley (1964), with emendations. Many treatments consider this taxon a subfamily of the Muscicapidae. In this checklist the groups traditionally placed in subfamilies of the Muscicapidae are treated as full families. Note that *Amalocichla* and *Drymodes* have been placed in the Eopsaltriidae following Schodde in Sibley & Ahlquist (1982a).

395. **Erithacus calliope** Siberian Rubythroat

Motacilla Calliope Pallas, 1776, Reise Versch. Prov. Russ. Reichs. 3: 697.

First recorded in region by Finch (1983c).

(*Luscinia calliope*, Rubythroat)

396. **Saxicola caprata** Pied Chat

Motacilla Caprata Linnaeus, 1766, Syst. Nat. ed. 12, 1: 335.

(Pied Bush Chat, Pied Stone Chat)

397. **Zoothera dauma** Scaly Thrush

Turdus Dauma Latham, 1790, Index Ornith. 1: 362.

Ford (1983) considered the population inhabiting the New Guinean region to constitute a species (*papuensis*) distinct from *dauma*. See White & Bruce (1985).

(*Oreocincla dauma*, White's Thrush, Scaled or Australian Ground Thrush, Golden Mountain Thrush, Groundthrush)

398. **Turdus poliocephalus** Island Thrush

Turdus poliocephalus Latham, 1801, Index Ornith., Suppl.: 44.

FAMILY **ORTHONYCHIDAE:** LOGRUNNERS AND ALLIES

Treatment follows Deignan (1964) in Peters' Checklist. Many of the species included were formerly placed in the Timaliidae. See also Sibley & Ahlquist (in press) for another view.

399. **Orthonyx temminckii** Logrunner

Orthonyx temminckii Ranzani, 1822, Elem. Zool. 3: 19.

(Brown or Southern Logrunner, Spine-tailed Chowchilla)

400. **Androphobus viridis** Papuan Whipbird

Androphilus viridus Rothschild and Hartert, 1911, Bull. Brit. Orn. Cl. 29: 33.

(Green-backed Babbler, Green-backed Shywit)

401. **Cinclosoma ajax** Painted Quail-thrush

Eupetes ajax Temminck, 1835, Pl. Col., livr. 97: pl. 573.

(Ajax Scrub Robin)

402. **Ptilorrhoa leucosticta** Spotted Jewel-babbler

Eupetes leucostictus Sclater, 1873 (1874), Proc. Zool. Soc. London: 690, pl. 52.

We follow Peters (1940) in placing this and the following two species in *Ptilorrhoa*.

(*Eupetes leucostictus*, High Mountain Eupetes)

403. **Ptilorrhoa caerulescens** Blue Jewel-babbler

Eupetes caerulescens Temminck, 1835, Pl. Col., livr. 97: pl. 574.

(*Eupetes caerulescens*, Lowland Eupetes)

404. **Ptilorrhoa castanonotus** Chestnut-backed Jewel-babbler

Eupetes castanonotus Salvadori, 1875, Ann. Mus. Civ. Stor. nat. Genova 7: 966.

The specific epithet is a noun in apposition, thus there is no change in gender with generic reassignment.

(*Ptilorrhoa castanonota*, Mid-mountain Eupetes)

405. **Melampitta lugubris** Lesser Melampitta

Melampitta lugubris 'Rosenberg' Schlegel, 1871 (1873), Nederl. Tijdsch. Dierk. 4: 47.

(Lesser Blackwit)

406. **Melampitta gigantea** Greater Melampitta

Mellopitta gigantea Rothschild, 1899, Ornith. Monatsber. 7: 137.

(Greater Blackwit)

407. **Ifrita kowaldi** Blue-capped Ifrita

Todopsis kowaldi De Vis, 1890, Ann. Rept. Brit. New Guinea, 1888-1889: 59.

(Blue-capped Babbler or Ifrit)

FAMILY **POMATOSTOMATIDAE:** FALSE BABBLERS

Formerly treated as true babblers (Deignan 1964), this group is now considered distinct by Sibley & Ahlquist (in press) who consider it a primitive member of the Australian 'corvine assemblage.'

408. **Pomatostomus isidorei** Rufous Babbler

Pomatostomus Isidorei Lesson, 1827, Dict. Sci. Nat., éd. Levrault 50: 37

Retention of this species in *Pomatostomus* follows Bell (1982).

(*Garritornis isidori*)

409. **Pomatostomus temporalis**
Grey-crowned Babbler

[Pomatorhinus] Temporalis Vigors and Horsfield, 1826 (1827), Trans. Linn. Soc. London 15: 330.

FAMILY **SYLVIIDAE**: OLD WORLD WARBLERS

Treatment follows Mayr (1941).

410. **Locustella fasciolata**
Gray's Grasshopper-Warbler

Acrocephalus fasciolatus G. R. Gray 1860, Proc. Zool. Soc. London: 349.

Recorded in eastern New Guinea by Finch (1982b).

(Grey-breasted Grasshopper Warbler, Grasshopper Warbler)

411. **Acrocephalus stentoreus**
Clamorous Reed-Warbler

Curruca stentorea Hemprich and Ehrenberg, 1833, Symb. Phys. Avium, fol. bb.

Earlier lists treated this form as conspecific with The Great Reed-Warbler *A. arundinaceus*. See Finch (1980b). The Australasian form is considered by some authorities to constitute a separate species, *A. australis* (White & Bruce 1985).

(*A. australis* or *A. arundinaceus australis*)

412. **Acrocephalus arundinaceus**
Great Reed-Warbler

Turdus arundinaceus Linnaeus, 1758, Syst. Nat. ed. 10, 1: 170.

The widespread resident population in New Guinea is now considered to be *A. stentoreus*, while *arundinaceus* occurs only as a rare migrant (LeCroy 1969, Finch 1980b).

(Greater Reedwarbler, Reed-Warbler)

413. **Megalurus timoriensis**
Tawny Grassbird

Megalurus timoriensis Wallace, 1863 (1864), Proc. Zool. Soc. London: 489.

(Rufous-capped Grass Warbler)

414. **Megalurus albolimbatus**
Fly River Grassbird

Poodytes albo-limbatus D'Albertis and Salvadori, 1879, Ann. Mus. Civ. Stor. nat. Genova 14: 87.

(D'Albertis Grassbird, Fly River Grass Warbler)

415. **Megalurus gramineus** Little Grassbird

Sphenoeacus gramineus Gould, 1845, Proc. Zool. Soc. London: 19.

(Little Grass Warbler)

416. **Cisticola exilis** Golden-headed Cisticola

[Malurus] exilis Vigors and Horsfield, 1827, ex Latham MS, Trans. Linn. Soc. London 15: 223.

(Fantail Warbler, Golden-headed Fantail Warbler)

417. **Cisticola juncidis** Zitting Cisticola

Sylvia Juncidis Rafinesque, 1809 (1810), Caratteri Alc. Nuovi Gen. Spec. Anim. Sicilia (1): 6 (spec. 10).

Added to the New Guinean list from sight records and specimens taken in the Bensbach region of the Western Province (Finch 1980d).

(Streaked Cisticola)

418. **Phylloscopus trivirgatus**
Island Leaf-Warbler

Phylloscopus trivirgatus Strickland, 1849, Contr. Ornith. 1: 123, pl. 34, fig. 2.

See White & Bruce (1985) for a different interpretation of the Indonesian populations of *Phylloscopus*.

(Leaf Warbler, Mountain Leaf Warbler)

FAMILY **MALURIDAE:** FAIRY-WRENS

Treatment follows Schodde & Weatherly (1982a) with emendations.

419. **Sipodotus wallacii** Wallace's Fairy-wren

Todopsis wallacii G. R. Gray, 1861 (1862), Proc. Zool. Soc. London: 429, 434, pl. 43, fig. 2.

(Blue-capped Wren, Wallace's Wren, Wallace's Wren Warbler)

420. **Malurus grayi** Broad-billed Fairy-wren

Todopsis grayi Wallace, 1862, Proc. Zool. Soc. London: 166.

We follow Diamond & LeCroy (in prep.), in treating the distinctive population *campbelli* (Schodde & Weatherly 1982a, b) as a subspecies of *M. grayi*. Note that because of timing of publications, the holotype for this form is not a photograph (cf. Schodde & Weatherly 1982b) but is, instead, a painting by Weatherly that appears in Schodde & Weatherly (1982a).

(*Chenorhamphus greyi, Malurus campbelli*, Broad-billed Wren, Broad-billed Wren Warbler, Campbell's Fairy-wren)

421. **Malurus cyanocephalus** Emperor Fairy-wren

Todus cyanocephalus Quoy and Gaimard, 1830, Voy. Astrolabe, Zool. 1: 227, pl. 5, fig. 4.

(*Todopsis cyanocephalus, Todopsis cyanocephala*, Blue Wren Warbler, Imperial Wren)

422. **Malurus alboscapulatus** White-shouldered Fairy-wren

Malurus alboscapulatus A. B. Meyer, 1874, Sitzungsb. Akad. Wiss. Wien 69: 496.

(Black and White Wren Warbler, White-shouldered Wren)

423. **Clytomyias insignis** Orange-crowned Fairy-wren

Clytomyias insignis Sharpe, 1879, Notes Leyden Mus. 1 (10): 31.

(Rufous Wren Warbler)

FAMILY **ACANTHIZIDAE:** SCRUBWRENS, THORNBILLS, AUSTRALIAN WARBLERS AND ALLIES

Treatment of this group as a distinct family follows Schodde (1975) and others. The sequence of genera and species follows Mayr (1941). Inclusion of *Crateroscelis* in this family follows Schodde (1975) and Mayr (1985)

424. **Crateroscelis murina** Rusty Mouse-Warbler

Brachypteryx murinus Sclater, 1858, Journ. Linn. Soc. London, Zool. 2: 158.

(Lowland Mouse-babbler, Lowland Mouse Warbler, Chanting Scrubwren or Fernwren)

425. **Crateroscelis nigrorufa** Bicoloured Mouse-warbler.

Sericornis nigro-rufa Salvadori, 1894, Ann. Mus. Civ. Stor. nat. Genova, ser. 2, 14: 151.

(Black-headed Scrubwren, Mid-mountain Mouse-babbler).

426. **Crateroscelis robusta** Mountain Mouse-Warbler

Gerygone robusta De Vis, 1898, Ann. Rept. Brit. New Guinea, 1896-1897: 84.

(Mountain Mouse-babbler, White-throated Scrubwren)

427 **Sericornis spilodera** Pale-billed Scrub-wren

Entomophila? spilodera G. R. Gray, 1859, Proc. Zool. Soc. London: 155.

(Pale-billed Sericornis)

428. **Sericornis beccarii** Beccari's Scrub-wren

Sericornis beccarii Salvadori, 1874, Ann. Mus. Civ. Stor. nat. Genova 6: 79.

(Beccari's Sericornis, Tropical Scrub-wren, Little Scrubwren)

429. **Sericornis virgatus**
Perplexing Scrub-wren

Crateroscelis virgata Reichenow, 1915, Journ. f. Ornith. 63: 128.

There are a number of obscure Papuan populations of *Sericornis* that have been placed variously into the species *beccarii, virgatus* and *nouhuysi*. We follow the treatment of Diamond (1969). See also Gilliard & LeCroy (1970).

(Streaked Scrubwren, Perplexing Sericornis)

430. **Sericornis nouhuysi** Large Scrub-wren

Sericornis arfakiana nouhuysi van Oort, 1909, Nova Guinea 9, Zool. (1): 90.

(Noisy Scrubwren, Large Mountain Sericornis)

431. **Sericornis perspicillatus**
Buff-faced Scrub-wren

Sericornis perspicillata Salvadori, 1896, Ann. Mus. Civ. Stor. nat. Genova ser. 2, 16: 99.

Considered by Mayr (1941) to be a subspecies of *S. rufescens*.

(Buffy-faced Scrubwren, Buff-faced Sericornis)

Note: The type of *Sericornis nigroviridis* Miller, 1964, is now considered to be a melanistic individual of *perspicillatus* (Beehler 1978).

432. **Sericornis rufescens**
Vogelkop Scrub-wren

Gerygone ? rufescens Salvadori, 1875, Ann. Mus. Civ. Stor. nat. Giacomo Daria 7: 961.

(Arfak Buff-faced Sericornis, Rufous Scrubwren)

433. **Sericornis papuensis**
Papuan Scrub-wren

Acanthiza papuensis De Vis, 1894, Ann. Rept. Brit. New Guinea, 1893-1894: 102.

(Papuan Sericornis, Olive Scrubwren)

434. **Sericornis arfakianus**
Grey-green Scrub-wren

Gerygone ? arfakiana Salvadori, 1875 (1876), Ann. Mus. Civ. Stor. nat. Genova 7: 960.

(Dusky Scrubwren, Grey-green Sericornis)

435. **Acanthiza murina**
New Guinea Thornbill

Gerygone murina De Vis, 1897, Ibis: 377.

(Mountain Thornbill, De Vis Tree Warbler)

436. **Gerygone cinerea** Grey Gerygone

Gerygone cinerea Salvadori, 1875, Ann. Mus. Civ. Stor. nat. Genova 7: 958.

(Grey Warbler, Grey Gerygone Warbler)

437. **Gerygone chrysogaster** Yellow-bellied Gerygone

Gerygone chrysogaster G. R. Gray, 1858, Proc. Zool. Soc. London: 174.

(Yellow-bellied Warbler, Yellow-bellied Gerygone Warbler)

438. **Gerygone chloronotus** Green-backed Gerygone

Gerygone chloronotus Gould, 1842 (1843), Proc. Zool. Soc. London: 133.

The specific epithet is here considered to be a masculine noun in apposition and thus is not altered to agree in gender with its feminine genus (*contra* Mayr 1941, *et mult.*).

(*Gerygone chloronota*, Green-backed Warbler, Grey-headed Gerygone Warbler)

439. **Gerygone palpebrosa** Fairy Gerygone

Gerygone palpebrosa Wallace, 1865, Proc. Zool. Soc. London: 475.

(Black-headed Gerygone Warbler, Fairy Warbler)

440. **Gerygone magnirostris**
Large-billed Gerygone

Gerygone magnirostris Gould, 1842 (1843), Proc. Zool. Soc. London: 133.

(Large-billed Warbler, Swamp Gerygone Warbler)

441. **Gerygone levigaster**
Mangrove Gerygone

Gerygone levigaster Gould, 1842 (1843), Proc. Zool. Soc. London: 133.

Formerly considered a population of the widespread *G. fusca* (Mayr 1941); our treatment follows Schodde (1975) and Ford (1981a).

(*Gerygone laevigaster*, Western Gerygone Warbler, Mangrove Warbler, Western Gerygone)

442. **Gerygone ruficollis** Brown-breasted Gerygone

Gerygone ruficollis Salvadori, 1875, Ann. Mus. Civ. Stor. Nat. Genova 7: 959.

(Treefern Gerygone Warbler, Brown-breasted Warbler)

443. **Gerygone olivacea** White-throated Gerygone

Psilopus olivaceus Gould, 1838, Synops Bds. Austr. 4: pl. 61.

(White-throated Warbler, White-throated Gerygone Warbler)

FAMILY **MUSCICAPIDAE:** OLD WORLD FLYCATCHERS

Treatment follows Mayr (1941).

444. **Muscicapa griseisticta** Grey-streaked Flycatcher

Hemichelidon griseisticta Swinhoe, 1861, Ibis: 330.

(Grey-spotted Flycatcher)

FAMILY **RHIPIDURIDAE:** FANTAILS

Familial assignment and treatment follows Sibley (1970), Voous (1977) and Mayr (1985). Sequence of species follows Mayr (1985).

445. **Rhipidura threnothorax** Sooty Thicket-Fantail

Rhipidura threnothorax S. Müller, 1843, Verh. Nat. Gesch., Land-en Volkenkunde 1: 185 (note).

446. **Rhipidura maculipectus** Black Thicket-Fantail

Rhipidura maculipectus G. R. Gray, 1858, Proc. Zool. Soc. London: 176.

447. **Rhipidura leucothorax** White-bellied Thicket-Fantail

Rhipidura leucothorax Salvadori, 1874, Ann. Mus. Civ. Stor. nat. Genova 6: 311.

(Black-throated Thicket Fantail, White-breasted Thicket Fantail)

448. **Rhipidura rufifrons** Rufous Fantail

Muscicapa rufifrons Latham, 1801, Index Ornith., Suppl: 1.

(Black-breasted Rufous Fantail)

449. **Rhipidura rufidorsa** Rufous-backed Fantail

Rhipidura rufidorsa A. B. Meyer, 1874, Sitzungsb. Akad. Wiss. Wien 70: 200.

(Grey-breasted Rufous Fantail)

450. **Rhipidura brachyrhyncha** Dimorphic Fantail

Rhipidura brachyrhyncha Schlegel, 1871 (1873), Nederl. Tijdschr. Dierk. 4: 42.

(Dimorphic Rufous Fantail)

451. **Rhipidura phasiana** Mangrove Fantail

Rhipidura phasiana De Vis, 1884, Proc. Roy. Soc. Queensland 1: 156.

Our treatment follows Ford (1981b), who considered *fuliginosa* and *phasiana* as specifically distinct. Most, or all, records of the 'Grey Fantail' in New Guinea are referable to *phasiana*, although *R. fuliginosa* may occur in the Fly River region as a migrant from Australia.

452. **Rhipidura atra** Black Fantail

Rhipidura atra Salvadori, 1875, Ann. Mus. Civ. Stor. nat. Genova 7: 922.

453. **Rhipidura hyperythra** Chestnut-bellied Fantail

Rhipidura hyperythra G. R. Gray, 1858, Proc. Zool. Soc. London: 176.

454. **Rhipidura albolimbata** Friendly Fantail

Rhipidura albolimbata Salvadori, 1874, Ann. Mus. Civ. Stor. nat. Genova 6: 312.
(White-eared Fantail)

455. **Rhipidura rufiventris** Northern Fantail
Platyrhynchus rufiventris Vieillot, 1818, Nouv. Dict. Hist. Nat., nouv. éd. 27: 21.
(White-throated Fantail)

456. **Rhipidura leucophrys** Willie Wagtail
Turdus leucophrys Latham, 1801, Index Ornith., Suppl.: 45.

FAMILY **MYIAGRIDAE:** MONARCHS

Elevation of this group to familial status follows Mayr (1985). Stem nomenclature (*Myiagr-*) follows Boles (1981). Sequence follows Mayr (1985) with emendations. *Machaerirhynchus* is tentatively included with this family, following Mayr (1985). Following Sibley & Ahlquist (1984), *Peltops* is transferred to the Cracticidae.

457. **Monarcha axillaris** Black Monarch
Monarcha axillaris Salvadori, 1875, Ann. Mus. Civ. Stor. nat. Genova 7: 921.
(Black Monarch Flycatcher, Fantail Monarch; note that Peckover & Filewood (1976) use the name 'Black Monarch' for *Myiagra atra*)

458. **Monarcha rubiensis** Rufous Monarch
Tchitrea rubiensis A. B. Meyer, 1874, Sitzungsb. Akad. Wiss. Wien 69: 494.
(Rufous Monarch Flycatcher)

459. **Monarcha cinerascens** Island Monarch
Drymophila cinerascens Temminck 1827, Pl. Col., livr. 72: pl. 430, fig. 2.
(Islet Monarch, Island Grey-headed Monarch Flycatcher)

460. **Monarcha melanopsis**
Black-faced Monarch
Muscicapa melanopsis Vieillot, 1818, Nouv. Dict. Hist. Nat., nouv. éd. 21: 450.
(Grey-winged Monarch Flycatcher, Black-faced Flycatcher)

461. **Monarcha frater**
Black-winged Monarch
Monarcha frater Sclater, 1873 (1874), Proc. Zool. Soc. London: 691
(Black-winged Monarch Flycatcher, Pearly or Black-chinned Flycatcher)

462. **Monarcha guttula**
Spot-winged Monarch
Muscicapa guttula Garnot, 1829, Voy. Coquille, Atlas: pl. 16, fig. 2.
The genus *Monarcha* has been shown to be masculine (Schodde 1977, Mees 1982b); however, the specific epithet *guttula* is a noun in apposition (feminine).
(Spot-winged Monarch Flycatcher)

463. **Monarcha trivirgatus**
Spectacled Monarch
Drymophila trivirgata Temminck, 1826, Pl. Col., livr. 70: pl. 418, fig. 1.
(*Monarcha trivirgata*, Spectacled Flycatcher)

464. **Monarcha julianae** Kofiau Monarch
Monarcha julianae Ripley, 1959, Postilla 38: 9.
(Kofiau Monarch Flycatcher)

465. **Monarcha manadensis**
Hooded Monarch
Muscicapa manadensis Quoy and Gaimard, 1830, Voy. Astrolabe, Zool. 1: 174.
(Black and White Monarch Flycatcher, White-bellied Monarch)

466. **Monarcha brehmii** Biak Monarch
Monarcha Brehmii Schlegel, 1871 (1873), Nederl. Tijdschr. Dierk. 4: 14.
(Biak Monarch Flycatcher)

467. **Monarcha chrysomela**
Golden Monarch
Muscicapa chrysomela Lesson and Garnot, 1828, Voy. Coquille, Atlas: pl. 18, fig. 2.
(Black and Yellow Monarch Flycatcher)

468. **Arses telescophthalmus**
Frilled Monarch

Muscicapa telescophthalmus Lesson, 1828, Man. Ornith. 1: 190.

Schodde & Weatherly (1982a) suggested that the population *insularis* may be specifically distinct from *telescophthalmus*.

(*Monarcha telescophthalmus*, *Arses lorealis*, Frilled or Frill-necked Flycatcher)

469. **Myiagra atra** Biak Black Flycatcher

Myiagra atra A. B. Meyer, 1874, Sitzungsb. Akad. Wiss. Wien 69: 498.

(Black Myiagra Flycatcher, Black Monarch)

470. **Myiagra rubecula** Leaden Flycatcher

Todus rubecula Latham, 1801, Index Ornith., Suppl.: 32.

(Leaden Monarch, Leaden Myiagra Flycatcher)

471. **Myiagra ruficollis**
Broad-billed Flycatcher

Platyrhynchos ruficollis Vieillot, 1818, Nouv. Dict. Hist. Nat., nouv. éd. 27: 13.

(Broad-billed Myiagra Flycatcher, Broad-billed Monarch)

472. **Myiagra cyanoleuca** Satin Flycatcher

Platyrhynchos cyanoleucus Vieillot, 1881, Nouv. Dict. Hist. Nat., nouv. éd. 27: 11.

(Satin Monarch, Satin Myiagra Flycatcher)

473. **Myiagra alecto** Shining Flycatcher

Drymophila alecto Temminck, 1827, Pl. Col., livr. 72: pl. 430, fig. 1.

Inclusion of this and the following species in *Myiagra* follows Schodde & McKean (1976), but for another view see Mees (1982b).

(*Monarcha alecto*, *Piezorhynchus alecto*, Shining Monarch Flycatcher)

474. **Myiagra inquieta** Restless Flycatcher

Turdus inquietus Latham, 1801, Index Ornith., Suppl.: 40.

(*Seisura inquieta*, *Monarcha inquieta*, Restless Monarch)

475. **Machaerirhynchus flaviventer**
Yellow-breasted Boatbill

Machaerirhynchus flaviventer Gould, 1850 (1852), Proc. Zool. Soc. London: 277, pl. 33.

(Yellow-breasted Flatbill Flycatcher, Yellow-breasted Flatbill, Boat-billed Flycatcher)

476. **Machaerirhynchus nigripectus**
Black-breasted Boatbill

Machaerirhynchus nigripectus Schlegel, 1871 (1873), Nederl. Tijdschr. Dierk. 4: 43.

(Black-breasted Flatbill, Black-breasted Flatbill Flycatcher)

FAMILY **EOPSALTRIIDAE:** AUSTRALIAN ROBINS

Treatment follows Mayr (1985), with emendations.

477. **Monachella muelleriana**
Torrent Flycatcher

Muscicapa Mülleriana Schlegel, 1871 (1873), Nederl. Tidjschr. Dierk. 4: 40.

(*Monachella mulleriana*, River Flycatcher)

478. **Microeca leucophaea** Jacky Winter

Sylvia leucophaea Latham, 1801, Index Ornith., Suppl.: 55.

(White-tailed or Brown Flycatcher)

479. **Microeca flavigaster**
Lemon-bellied Flycatcher

Microeca flavigaster Gould, 1842 (1843), Proc. Zool. Soc. London: 132.

(Lemon-breasted or Yellow-bellied Flycatcher)

480. **Microeca griseoceps**
Yellow-legged Flycatcher

Microeca griseoceps De Vis, 1894, Ann. Rept. Brit. New Guinea, 1893-1894: 101.

(Grey-headed, Yellow-footed or Yellow Flycatcher, Yellow-footed Microeca Flycatcher)

481. **Microeca flavovirescens**
Olive Flycatcher

Microeca flavovirescens G. R. Gray, 1858, Proc. Zool. Soc. London: 178.

(Orange-chinned Flycatcher, Olive Microeca Flycatcher)

482. **Microeca papuana** Canary Flycatcher

Microeca papuana A. B. Meyer, 1875, Sitzungsber. Abh. Nat. Ges. Isis, Dresden: 75.

(Papuan or Yellow Microeca Flycatcher)

483. **Tregellasia leucops** White-faced Robin

Leucophantes leucops Salvadori, 1875, Ann. Mus. Civ. Stor. nat. Genova 7: 921.

Placement of *Tregellasia* after *Microeca* follows Schodde (1975).

(*Eopsaltria leucops*, White-faced Flycatcher)

484. **Eugerygone rubra** Garnet Robin

Pseudogerygone rubra Sharpe, 1878, Notes Leyden Mus. 1: 29.

In earlier lists this genus was often placed with the warblers; recent studies (cf. Keast 1977, Orenstein 1979) argued that it should be included with the robins.

(Red-backed Warbler, Red-backed Flycatcher)

485. **Petroica bivittata** Mountain Robin

Petroeca bivitatta De Vis, 1897, Ibis: 376.

(Alpine Robin, Forest Robin Flycatcher)

486. **Petroica archboldi**
Snow Mountain Robin

Petroica archboldi Rand, 1940, Am. Mus. Novit. 1072: 5.

(Rock Robin Flycatcher, Snow Robin)

487. **Eopsaltria pulverulenta**
Mangrove Robin

Myiolestes pulverulentus Bonaparte, 1850, Consp. Gen. Av. 1: 358.

Generic designation follows Schodde (1975). See also Schodde (1981) for further discussion on relationships.

(*Poecilodryas pulverulenta*, *Penoenanthe pulverulenta*, Mangrove Flycatcher or Flyrobin)

488. **Poecilodryas albispecularis**
Ashy Robin

Pachycephala albispecularis Salvadori, 1875, Ann. Mus. Civ. Stor. nat. Genova 7: 931.

Generic designation follows Schodde (1975).

(*Heteromyias albispecularis*, Ground Thicket Flycatcher, White-browed Flyrobin, Grey-headed Robin)

489. **Poecilodryas brachyura** Black-chinned Robin

Leucophantes brachyurus Sclater, 1873 (1874), Proc. Zool. Soc. London: 691.

(White-breasted Flycatcher, White-breasted Flyrobin)

490. **Poecilodryas hypoleuca** Black-sided Robin

Petroica hypoleuca G. R. Gray, 1857, Proc. Zool. Soc. London: 155.

(Pied Flyrobin, Black and White Flycatcher)

491. **Poecilodryas albonotata** Black-throated Robin

Megalestes albonotata Salvadori, 1875, Ann. Mus. Civ. Stor. nat. Genova 7: 770.

(Black-throated Flycatcher, Black-throated Flyrobin)

492. **Poecilodryas placens** Banded Yellow Robin

Eopsaltria placens Ramsay, 1879, Proc. Linn. Soc. New South Wales 3: 272.

(Yellow Flyrobin, Olive-yellow Flycatcher)

493. **Drymodes superciliaris**
Northern Scrub-Robin

Drymodes superciliaris Gould, 1850, in Jardine's Contrib. Ornith.: 105.

Placement of *Drymodes* into the Eopsaltriidae follows Sibley & Ahlquist (1982a).

(Scrub Robin)

494. **Amalocichla sclateriana**
Greater Ground-Robin

Amalocichla sclateriana De Vis, 1892, Ann. Rept. Brit. New Guinea, 1890-1891: 95.

Inclusion of *Amalocichla* in the Eopsaltriidae, near *Drymodes* follows the suggestion of Schodde *in* Sibley & Ahlquist (1982a).

(Greater New Guinea Thrush, Cloud-forest Velvetthrush)

495. **Amalocichla incerta**
Lesser Ground-Robin

Eupetes incertus Salvadori, 1875, Ann. Mus. Civ. Stor. nat. Genova 7: 967.

(Lesser New Guinea Thrush, Rusty Velvet-thrush)

496. **Peneothello sigillatus**
White-winged Robin

Poecilodryas ? sigillata De Vis, 1890, Ann. Rept. Brit. New Guinea, 1888-1889: 59.

(White-winged Thicket-flycatcher, Black Flyrobin)

497. **Peneothello bimaculatus**
White-rumped Robin

Myiolestes ? bimaculatus Salvadori, 1875, Ann. Mus. Civ. Stor. nat. Genova 6: 84.

(White-rumped Thicket-flycatcher, White-rumped Flyrobin)

498. **Peneothello cryptoleucus**
Smoky Robin

Poecilodryas cryptoleucus Hartert, 1930, Novit. Zool. 36: 67.

(Grey Thicket-flycatcher, Smoky Flyrobin)

499. **Penoethello cyanus** Blue-grey Robin

Myiolestes ? cyanus Salvadori, 1874, Ann. Mus. Civ. Stor. nat. Genova 6: 84.

(Blue Flyrobin, Slaty Thicket-flycatcher)

500. **Pachycephalopsis hattamensis**
Green-backed Robin

Pachycephala hattamensis A. B. Meyer, 1874, Sitzungsb. Akad. Wiss. Wien 69: 391.

(Green-backed Whistler Robin, Green Thicket-flycatcher)

501. **Pachycephalopsis poliosoma**
White-eyed Robin

Pachycephala poliosoma Sharpe, 1882, Journ. Linn. Soc. London, Zool. 16: 318.

(White-throated Thicket-flycatcher, Grey-backed Whistler Robin)

FAMILY **PACHYCEPHALIDAE:** WHISTLERS, PITOHUIS AND ALLIES

Treatment follows Mayr (1941, 1967), with emendations.

502. **Pachycare flavogrisea** Dwarf Whistler

Pachycephala flavogrisea Meyer, 1874, Sitzungsb. Akad. Wiss. Wien 69: 495.

(Golden-faced Pachycare, Goldenface Whistler)

503. **Rhagologus leucostigma**
Mottled Whistler

Pachycephala leucostigma Salvadori, 1875, Ann. Mus. Civ. Stor. nat. Genova 7: 933.

504. **Pachycephala pectoralis**
Golden Whistler

M[uscicapa] pectoralis Latham, 1801, Index Ornith., Suppl.: 51.

Previous lists recognized a single species in the *pectoralis-melanura* complex in the New Guinean region. At present, we follow Galbraith (1967) in recognizing two. The following Papuan races included in *pectoralis: balim, collaris, misimae*, and *rosseliana*.

(Golden Whistler)

505. **Pachycephala melanura**
Mangrove Golden Whistler

Pachycephala melanura Gould, 1842 (1843), Proc. Zool. Soc. London: 134.

In earlier lists this complex was included in *pectoralis*. We follow Galbraith (1967) and subsequent authors in treating this as a full species. The populations *fergussonis* and *spinicauda* represent this species in the New Guinean region.

(Black-tailed or Robust Whistler)

506. **Pachycephala soror** Sclater's Whistler

Pachycephala soror Sclater, 1873 (1874), Proc. Zool. Soc. London: 692.

507. **Pachycephala schlegelii** Regent Whistler

Pachycephala Schlegelii Schlegel (ex Rosenberg MS), 1871 (1873), Nederl. Tijdschr. Dierk.: 43.

(Schlegel's Whistler)

508. **Pachycephala aurea** Golden-backed Whistler

Pachycephala aurea Reichenow, 1899, Ornith. Monatsb. 7: 131.

(Yellow-backed Whistler)

509. **Pachycephala lorentzi** Lorentz's Whistler

Pachycephala schlegelii lorentzi Mayr, 1931, Mitt. Zool. Mus. Berlin 17: 673.

510. **Pachycephala meyeri** Vogelkop Whistler

Pachycephala meyeri Salvadori, 1890, Agg. Ornith. Papuasia e Mol.: 104.

(Grey-crowned Whistler)

511. **Pachycephala simplex** Grey Whistler

Pachycephala simplex Gould, 1842 (1843), Proc. Zool. Soc. London: 135.

We follow Schodde (1975) in considering the Papuan populations a part of the *simplex* group, which includes subspecies formerly assigned to *griseiceps*.

(*Pachycephala griseiceps*, Grey-headed or Brown Whistler)

512. **Pachycephala hyperythra** Rusty Whistler

Pachycephala hyperythra Salvadori, 1875, Ann. Mus. Civ. Stor. nat. Genova 7: 932.

(Brownish Whistler, Rufous-breasted Whistler)

513. **Pachycephala phaionotus** Island Whistler

Myiolestes phaionotus Bonaparte, 1850, Consp. Gen. Av. 1: 358.

Since the specific epithet is here regarded as a noun in apposition, its ending should remain masculine, *contra* Mayr (1941, 1967) and Filewood & Peckover (1978).

(*Pachycephala phaionota, P. phaionotum*)

514. **Pachycephala modesta** Brown-backed Whistler

Poecilodryas modesta De Vis, 1894, Ann. Rept. Brit. New Guinea, 1893-1894: 101.

(Modest Whistler)

515. **Pachycephala monacha** Black-headed Whistler

Pachycephala monacha Gray, 1858, Proc. Zool. Soc. London: 179.

Following Mayr (1941) and Rand & Gilliard (1967), we treat this taxon as a full species, distinct from *rufiventris* (*contra* Mayr 1967)

516. **Pachycephala rufiventris** Rufous Whistler

Sylvia rufiventris Latham, 1801, Index Ornith., Suppl.: 54.

We follow Mayr (1941, 1967) in treating the Papuan populations *rufigaster* and *meeki* as members of the polytypic species *rufiventris*. Some workers (e.g. Mayr 1967) consider the Papuan populations of this and the preceding species to be conspecific.

(*P. leucogaster*, White-bellied Whistler)

517. **Pachycephala rufinucha** Rufous-naped Whistler

Pachycephala rufinucha Sclater, 1873 (1874), Proc. Zool. Soc. London: 692.

As pointed out by Schodde (*in litt.*), this species should perhaps be placed in the Eopsaltriidae near *Pachycephalopsis*.

(*Aleadryas rufinucha*)

518. **Pachycephala tenebrosa** Sooty Whistler

Pachycephala tenebrosa Rothschild, 1911, Bull. Brit. Orn. Cl. 29: 20.

519. **Colluricincla megarhyncha** Little Shrike-thrush

Muscicapa megarhyncha Quoy and Gaimard, 1830, Voy. Astrolabe, Zool. 1: 172, pl. 3, fig. 1.

Treatment follows Mayr (1944, 1967) and others.

(*Myiolestes megarhynchus*, Brown Shrike-flycatcher, Brown or Rufous Shrikethrush, Brown or Rufous Thrush)

520. **Colluricincla harmonica** Grey Shrike-thrush

Turdus harmonicus Latham, 1801, Index Ornith., Suppl.: 41.

(Grey Shrike-flycatcher or Thrush)

521. **Pitohui kirhocephalus** Variable Pitohui

Lanius kirhocephalus Lesson and Garnot, 1827, Voy. Coquille, Atlas: pl. 11; 1829, Zool. 1: 633.

522. **Pitohui dichrous** Hooded Pitohui

Rhectes dichrous Bonaparte, 1850, Compt. Rend. Acad. Sci., Paris 31: 563.

(Black-headed Pitohui)

523. **Pitohui incertus** White-bellied Pitohui

Pitohui incertus van Oort, 1909, Nova Guinea, Zool. 9: 94.

(Mottled or Mottle-breasted Pitohui)

524. **Pitohui ferrugineus** Rusty Pitohui

Rhectes ferrugineus Bonaparte, 1850, Compt. Rend. Acad. Sci., Paris 31: 563.

525. **Pitohui cristatus** Crested Pitohui

Rectes cristata Salvadori, 1875, Ann. Mus. Civ. Stor. nat. Genova 7: 930.

526. **Pitohui nigrescens** Black Pitohui

Rectes nigrescens Schlegel, 1871 (1873), Nederl. Tijdschr. Dierk. 4: 46.

(Dusky Pitohui)

527. **Eulacestoma nigropectus** Wattled Ploughbill

Eulacestoma nigropectus De Vis, 1894, Ann. Rept. Brit. New Guinea, 1893-1894: 102.

Some authorities (e.g. Rand & Gilliard 1967) place this and its relatives in the family Falcunculidae.

(Wattled Shrike-tit)

FAMILY **CLIMACTERIDAE:** AUSTRALIAN TREECREEPERS

Generic treatment follows Parker (1982) and Sibley *et al.* (1984). We follow Schodde (1976) and others in keeping the Papuan population specifically distinct. See also Boles & Longmore (1983).

528. **Cormobates placens** Papuan Treecreeper

Climacteris placens Sclater, 1873 (1874), Proc. Zool. Soc. London: 693.

Keast (1957) considered this population to be conspecific with *leucophaea* of Australia.

(*Climacteris placens, Climacteris leucophaea*, New Guinea Treecreeper, Papuan Creeper)

FAMILY **NEOSITTIDAE:** SITTELLAS

Treatment follows Schodde (1975). For affinities of this group, see Sibley & Ahlquist (1982b).

529. **Daphoenositta chrysoptera** Varied Sittella

Sitta chrysoptera Latham, 1801, Index Ornith., Suppl.: 32.

Earlier treatments gave the Papuan population full specific status (e.g. Mayr 1941, Rand & Gilliard 1967).

(*Neositta papuensis, Neositta chrysoptera*, Papuan Sittella, Mountain Sittella)

530. **Daphoenositta miranda** Black Sittella
Daphoenositta miranda De Vis, 1897, Ibis: 380.
(Pink-faced Nuthatch)

FAMILY **DICAEIDAE:** FLOWERPECKERS AND BERRYPECKERS

Treatment follows Salomonsen (1967).

531. **Melanocharis arfakiana** Obscure Berrypecker
Dicaeum arfakianum Finsch, 1900, Notes Leyden Mus. 22: 70.
Schodde (1978) notes that this form could be either a rare morph of *M. nigra* or a hybrid of *nigra* × *longicauda*, but see Opit (1975), Finch (1982a) for recent sight records.
(Plain Berrypecker)

532. **Melanocharis nigra** Black Berrypecker
Dicaeum niger Lesson, 1826 (1830), Voy. Coquille, Zool. 1: 673.

533. **Melanocharis longicauda** Mid-mountain Berrypecker
Melanocharis longicauda Salvadori, 1875 (1876), Ann. Mus. Civ. Stor. nat. Genova 7: 942.
(Yellow-bellied Berrypecker)

534. **Melanocharis versteri** Fan-tailed Berrypecker
Pristorhamphus versteri Finsch, 1875 (1876), Proc. Zool. Soc. London: 642.

535. **Melanocharis striativentris** Streaked Berrypecker
Melanocharis striativentris Salvadori, 1894, Ann. Mus. Civ. Stor. nat. Genova, ser. 2, 14: 150.
(*Neneba striativentris*)

536. **Rhamphocharis crassirostris** Spotted Berrypecker
Rhamphocharis crassirostris Salvadori, 1875 (1876), Ann. Mus. Civ. Stor. nat. Genova 7: 943.

537. **Dicaeum pectorale** Papuan Flowerpecker
Dicaeum pectorale S. Müller, 1843, Verh. Nat. Gesch., Land-en Volkenkunde: 162.
The species now includes the populations *geelvinkianum* and *nitidum* formerly considered distinct species (see Salomonsen 1967).
(Red-capped, Shining, Louisiade, or Olive-crowned Flowerpecker)

538. **Dicaeum hirundinaceum** Mistletoebird
Motacilla hirundinacea Shaw, 1792, Nat. Misc. 4: pl. 114.
(Mistletoe or Australian Flowerpecker)

539. **Oreocharis arfaki** Tit Berrypecker
Parus (?) Arfaki A. B. Meyer, 1875, Sitzungsb. Nat. Ges. Isis, Dresden: 75.
This and the following species are both considered to be aberrant bulbuls by Harrison & Parker (1966).
(Painted Berrypecker)

540. **Paramythia montium** Crested Berrypecker
Paramythia montium De Vis, 1892, Ann. Rept. Brit. New Guinea, 1890-1891: 95.

FAMILY **NECTARINIIDAE:** SUNBIRDS

Treatment follows Rand (1967), with emendation.

541. **Nectarinia aspasia** Black Sunbird
Cinnyris aspasia Lesson and Garnot 1828, Voy. Coquille, Atlas: pl. 30, fig. 4.
Mees (1966) noted that Hachisuka had determined that the specific name *sericea* is preoccupied, and that the next available name is *aspasia*.

(*Nectarinia sericea* or *sericeus, Cinnyrus sericeus*)

542. **Nectarinia jugularis**
Yellow-bellied Sunbird

Certhia jugularis Linnaeus, 1766, Syst. Nat., ed. 11: 185.

(*Cinnyris jugularis, Cyrtostomus frenatus*, Olive-backed, Black-throated or Yellow-breasted Sunbird)

FAMILY **ZOSTEROPIDAE:** WHITE-EYES

Treatment follows Mees (1969), as emended to treat *Zosterops* as masculine (Stoll *et al.* 1974).

543. **Zosterops chloris** Moluccan White-eye

Z[osterops] chloris Bonaparte, 1850, Consp. Gen. Av. 1: 398.

(Yellow White-eye, Mangrove White-eye)

544. **Zosterops atrifrons**
Black-fronted White-eye

Zosterops atrifrons Wallace, 1863 (1864), Proc. Zool. Soc. London: 493.

Includes *minor*, which has been considered a full species in some treatments (e.g. Rand & Gilliard 1967).

Note: *Zosterops citrinellus*, sometimes listed as *Z. palpebrosa* on earlier lists of the region, inhabits Australian islands in Torres Strait outside the range of the present work.

545. **Zosterops mysorensis** Biak White-eye

Zosterops mysorensis A. B. Meyer, 1874, Sitzungsb. Akad. Wiss. Wien 70: 116.

(Soepiori White-eye)

546. **Zosterops meeki** Tagula White-eye

Zosterops meeki Hartert, 1898, Novit. Zool. 5: 528.

(White-throated White-eye)

547. **Zosterops fuscicapillus**
Western Mountain White-eye

Zosterops fuscicapilla Salvadori, 1875, Ann. Mus. Civ. Stor. nat. Genova 7: 955.

(*Zosterops fuscicapilla*, Yellow-bellied Mountain White-eye, Yellow-bellied White-eye)

548. **Zosterops novaeguineae** New Guinea White-eye

Zosterops novaeguineae Salvadori, 1878, Ann. Mus. Civ. Stor. nat. Genova 12: 341.

(Mountain White-eye, New Guinea Mountain White-eye)

549. **Zosterops griseotinctus**
Louisiades White-eye

Zosterops griseotinctus G. R. Gray, 1858, Proc. Zool. Soc. London: 175.

(Island White-eye)

FAMILY **MELIPHAGIDAE:** HONEYEATERS

Treatment follows Salomonsen (1967), with emendations.

550. **Timeliopsis fulvigula** Olive Straightbill

Euthyrhynchus fulvigula Schlegel, 1871 (1873), Nederl. Tijdschr. Dierk. 4: 40.

(Olive or Mountain Straight-billed Honeyeater)

551. **Timeliopsis griseigula**
Tawny Straightbill

Euthyrhynchus griseigula Schlegel, 1871 (1873), Nederl. Tijdschr. Dierk. 4: 39.

(Babbler Honeyeater, Tawny or Lowland Straight-billed Honeyeater)

552. **Melilestes megarhynchus** Long-billed Honeyeater

Ptilotis megarhynchus G. R. Gray, 1858, Proc. Zool. Soc. London: 174.

553. **Toxorhamphus novaeguineae**
Yellow-bellied Longbill

Cinnyris Novae-Guineae Lesson, 1827, Dict. Sci. Nat, éd. Levrault 1: 22.

Placed in the Nectariniidae by Peckover & Filewood (1976).

(Canary Bowbill)

554. **Toxorhamphus poliopterus**
Slaty-chinned Longbill

Melilestes poliopterus Sharpe, 1882, Journ. Linn. Soc. London, Zool. 16: 318.

Placed in the Nectariniidae by Peckover & Filewood (1976).

(Grey-winged Bowbill)

555. **Oedistoma iliolophus**
Dwarf Honeyeater

Melilestes iliolophus Salvadori, 1875 (1876), Ann. Mus. Civ. Stor. nat. Genova 7: 951.

The specific epithet *iliolophus* is here considered a masculine noun in apposition the ending of which does not change to agree in gender with its genus (*contra* Salomonsen 1967).

(*Toxorhamphus iliolophus, Oedistoma iliolophum*, Grey-bellied Longbill, Grey-bellied Honeyeater)

556. **Oedistoma pygmaeum**
Pygmy Honeyeater

Oedistoma pygmaeum Salvadori, 1875 (1876) Ann. Mus. Civ. Stor. nat. Genova 7: 952.

557. **Glycichaera fallax**
Green-backed Honeyeater

Glycichaera fallax Salvadori, 1878, Ann. Mus. Civ. Stor. nat. Genova 12: 335.

(White-eyed or Pale-eyed Honeyeater)

558. **Lichmera argentauris**
Olive Honeyeater

Ptilotis argentauris Finsch, 1870, Abh. Nat. Verh. Bremen 2: 364.

(Plain Olive Honeyeater)

559. **Lichmera indistincta**
Brown Honeyeater

[Meliphaga] indistincta Vigors and Horsfield, 1827, Trans. Linn. Soc. London 15: 315.

560. **Lichmera alboauricularis** Silver-eared Honeyeater

Stigmatops albo-auricularis Ramsay, 1878 (1879), Proc. Linn. Soc. New South Wales 3: 75.

561. **Myzomela albigula**
White-chinned Myzomela

Myzomela albigula Hartert, 1898, Bull. Brit. Orn. Cl. 8: 20.

(White-throated Honeyeater)

562. **Myzomela eques**
Red-throated Myzomela

Cinnyris eques Lesson, 1827, Bull. Sci. Nat. Férussac 11: 386.

(Ruby-throated Honeyeater, Red-spot or Red-spotted Myzomela)

563. **Myzomela obscura** Dusky Myzomela

Myzomela obscura Gould, 1842 (1843), Proc. Zool. Soc. London: 136.

(Dusky Honeyeater)

564. **Myzomela cruentata** Red Myzomela

Myzomela cruentata A. B. Meyer, 1874 (1875), Sitzungsb. Akad. Wiss. Wien 70: 202.

(Red Honeyeater)

565. **Myzomela nigrita**
Papuan Black Myzomela

Myzomela nigrita G. R. Gray, 1858, Proc. Zool. Soc. London: 173.

(Black Myzomela, Carbon Honeyeater)

566. **Myzomela erythrocephala**
Red-headed Myzomela

Myzomela erythrocephala Gould, 1839 (1840), Proc. Zool. Soc. London: 144.

(Red-headed Honeyeater, Mangrove Red-headed Myzomela)

567. **Myzomela adolphinae**
Mountain Red-headed Myzomela

Myzomela adolphinae Salvadori, 1875 (1876), Ann. Mus. Civ. Stor. nat. Genova 7: 946.

(Elfin Honeyeater)

568. **Myzomela sclateri** Sclater's Myzomela

Myzomela sclateri Forbes, 1879, Proc. Zool. Soc. London: 265, pl. 25, fig. 2.

(Scarlet-throated Honeyeater)

569. **Myzomela rosenbergii**
Red-collared Myzomela

Myzomela rosenbergii Schlegel, 1871 (1873), Nederl. Tijdschr. Dierk. 4: 38.

(Black and Red Honeyeater, Red-caped Honeyeater)

570. **Meliphaga mimikae**
Spot-breasted Meliphaga

Ptilotis mimikae Ogilvie-Grant, 1911, Bull. Brit. Ornith. Cl. 29: 27.

(Greater or Large Spot-breasted Meliphaga)

571. **Meliphaga montana**
Forest White-eared Meliphaga

Ptilotis montana Salvadori, 1880, Ann. Mus. Civ. Stor. nat. Genova 16: 77.

Earlier treatments (*cf.* Rand 1936) included populations of *albonotata* with those of *montana*; the populations known as *auga* and *setekwa* belong in the species *albonotata*.

(White-eared Mountain Meliphaga, Black-fronted Honeyeater)

572. **Meliphaga orientalis**
Mountain Meliphaga

Ptilotis flavirictus orientalis A. B. Meyer, 1894, Journ. f. Ornith. 42: 92.

(Small or Lesser Spot-breasted Meliphaga)

573. **Meliphaga albonotata**
Scrub White-eared Meliphaga

Ptilotis albonotata Salvadori, 1876, Ann. Mus. Civ. Stor. nat. Genova 9: 33.

In many instances, previous treatments have inadvertently placed populations of this species in *montana. Meliphaga montana auga* Rand, sometimes treated as a full species (*cf.* Diamond 1972, Peckover & Filewood 1976) is a synonym of *albonotata*.

(White-marked Honeyeater, White-marked Meliphaga, Southern White-eared Meliphaga, Diamond Honeyeater)

574. **Meliphaga aruensis**
Puff-backed Meliphaga

Ptilotis aruensis Sharpe, 1884, Rept. Zool. Coll. Voy. Alert: 19.

(Puff-backed Honeyeater)

575. **Meliphaga analoga** Mimic Meliphaga

Ptilotis analoga Reichenbach, 1852, Handbook Spec. Ornith., Abth. 2, Meropinae: 103, tab. 467.

(Mimic Honeyeater)

576. **Meliphaga vicina** Tagula Meliphaga

Ptilotis analoga vicina Rothschild and Hartert, 1912, Novit. Zool. 19: 203.

(Louisiades Honeyeater)

577. **Meliphaga gracilis** Graceful Meliphaga

Ptilotis gracilis Gould, 1866, Proc. Zool. Soc. London: 217.

(Slender-billed Meliphaga, Graceful Honeyeater)

578. **Meliphaga flavirictus**
Yellow-gaped Meliphaga

Ptilotis flavirictus, Salvadori, 1880, Ann. Mus. Civ. Stor. nat. Genova 16: 76.

(Yellow-gaped Honeyeater)

579. **Lichenostomus versicolor**
Varied Honeyeater

Ptilotis versicolor Gould, 1842 (1843), Proc. Zool. Soc. London: 136.

Formerly considered conspecific with *virescens*. Ford (1978) discusses geographic variation and species limits among regional populations. We follow Schodde (1975) in placing this and the subsequent four species in the genus *Lichenostomus*.

(*Meliphaga versicolor*)

580. **Lichenostomus flavescens**
Yellow-tinted Honeyeater

Ptilotis flavescens Gould, 1839 (1840), Proc. Zool. Soc. London: 144.

Treatment follows Mayr (1941), *contra* Salomonsen (1967), who considered the New Guinean form to be conspecific with *L. fuscus*; see Finch (1983b) for notes questioning the affinities of the Papuan populations.

(*Meliphaga flavescens*, Yellowish Honeyeater)

581. **Lichenostomus subfrenatus** Black-throated Honeyeater

Ptilotis subfrenata Salvadori, 1875 (1876), Ann. Mus. Civ. Stor. nat. Genova 7: 948.

For notes on relationships see Longmore & Boles (1983).

(*Oreornis subfrenatus, Meliphaga subfrenata*)

582. **Lichenostomus obscurus** Obscure Honeyeater

Ptilotis obscura De Vis, 1897, Ibis: 383.

(*Oreornis obscurus, Meliphaga obscura*, Lemon-cheeked Honeyeater)

583. **Lichenostomus chrysogenys** Orange-cheeked Honeyeater

Oreornis chrysogenys van Oort, 1910, Notes Leyden Mus. 32: 215.

(*Meliphaga chrysogenys*)

584. **Xanthotis polygramma** Spotted Honeyeater

Ptilotis polygramma G. R. Gray, 1861, Proc. Zool. Soc. London: 429.

Placement of this and the following species in Xanthotis follows Mayr (1941).

(*Meliphaga polygramma*, Spotted Xanthotis, Many-spotted Honeyeater)

585. **Xanthotis flaviventer** Tawny-breasted Honeyeater

Myzantha flaviventer Lesson, 1828, Man. Ornith. 2: 67. New Name for *Philedon chrysotis* Latham, 1801 (indeterminable).

(*Xanthotis chrysotis, Meliphaga flaviventer* or *chrysotis*, Brown Xanthotis, Buff-breasted Honeyeater)

586. **Melithreptus albogularis** White-throated Honeyeater

Melithreptus albogularis Gould, 1848, Bds. Australia 4(30): pl. 74; also 1847 (1848), Proc. Zool. Soc. London 15: 220.

Rand & Gilliard (1967) treated the New Guinean population as synonymous with *M. lunatus*.

587. **Entomyzon cyanotis** Blue-faced Honeyeater

Gracula cyanotis Latham, 1801, Index Ornith., Suppl.: 29.

(Blue-cheeked Honeyeater)

588. **Pycnopygius ixoides** Plain Honeyeater

Ptilotis ? ixoides Salvadori, 1878, Ann. Mus. Civ. Stor. nat. Genova 12: 338.

(Brown Honeyeater, Nondescript Honeyeater)

589. **Pycnopygius cinereus** Marbled Honeyeater

Ptilotis cinerea Sclater, 1873 (1874), Proc. Zool. Soc. London: 693.

(Grey Honeyeater)

590. **Pycnopygius stictocephalus** Streak-headed Honeyeater

Pycnonotus (?) stictocephalus Salvadori, 1876, Ann. Mus. Civ. Stor. nat. Genova 9: 34.

(Streak-capped Honeyeater)

591. **Philemon meyeri** Meyer's Friarbird

Philemon meyeri Salvadori, 1878, Ann. Mus. Civ. Stor. nat. Genova 12: 339, new name for *Tropidorhynchus inornatus* A. B. Meyer, 1875, Sitzun sb. Akad. Wiss. Wien. 70: 212.

(Dwarf Friarbird)

592. **Philemon brassi** Brass's Friarbird

Philemon brassi Rand, 1940 Am. Mus. Novit. 1072: 13.

(Brass's Friar Bird)

593. **Philemon citreogularis** Little Friarbird

Tropidorhynchus citreogularis Gould, 1837 Synops. Bds. Aust. 1: pl. 113, fig. 1; also 1836 (1837), Proc. Zool. Soc. London: 143.

(Yellow-throated Friar Bird)

594. **Philemon buceroides**
Helmeted Friarbird

Philedon buceroides Swainson, 1838, Anim. Menag.: 325.

Treatment follows Schodde (1975).

(*Philemon novaeguineae*, New Guinea Friar Bird, Leatherhead)

595. **Philemon corniculatus** Noisy Friarbird

Merops corniculatus Latham, 1790, Index Ornith. 1: 276.

(Bald Friarbird)

596. **Ptiloprora plumbea**
Leaden Honeyeater

Ptilotis plumbea Salvadori, 1894, Ann. Mus. Civ. Stor. nat. Genova, ser. 2, 14: 151.

597. **Ptiloprora meekiana** Yellowish Streaked Honeyeater

Ptilotis meekiana Rothschild and Hartert, 1907, Novit. Zool. 14: 482.

(Meek's Streaked Honeyeater, Meek's Honeyeater)

598. **Ptiloprora erythropleura** Rufous-sided Honeyeater

Ptilotis erythropleura Salvadori, 1875 (1876), Ann. Mus. Civ. Stor. nat. Genova 7: 949.

(Red-sided Honeyeater, Red-sided Streaked Honeyeater)

599. **Ptiloprora guisei**
Rufous-backed Honeyeater

Ptilotis (?) guisei De Vis, 1894, Ann. Rept. Brit. New Guinea, 1893-1894: 103.

Diamond (1969) considered the northern, coastal populations *mayri* and *acrophila* as constituting a full species *P. mayri*.

(Red-backed, Green-eyed, or Guise's Honeyeater, Brown-backed Streaked Honeyeater)

600. **Ptiloprora perstriata**
Grey-streaked Honeyeater

Ptilotis perstriata De Vis, 1898, Ann. Rept. Brit. New Guinea, 1896-1897: 86

(Many-streaked Honeyeater, Black-backed Streaked Honeyeater)

601. **Melidectes fuscus** Sooty Melidectes

Acanthochoera fusca De Vis, 1897, Ibis: 383.

(Sooty Honeyeater)

602. **Melidectes princeps**
Long-bearded Melidectes

Melidectes princeps Mayr and Gilliard, 1951, Amer. Mus. Novit. 1524: 13.

(Long-bearded Honeyeater)

603. **Melidectes nouhuysi**
Short-bearded Melidectes

Melirrhophetes nouhuysi van Oort, 1910, Notes Leyden Mus. 32: 215.

(Short-bearded Honeyeater)

604. **Melidectes ochromelas**
Cinnamon-browed Melidectes

Melirrhophetes ochromelas A. B. Meyer, 1874 (1875), Sitzungsb. Akad. Wiss. Wien 70: 111.

(Mid-mountain Melidectes, Cinnamon-browed Honeyeater)

605. **Melidectes leucostephes**
Vogelkop Melidectes

Melirrhophetes leucostephes A. B. Meyer, 1874 (1875), Sitzungsb. Akad. Wiss. Wien 70: 110.

(White-fronted Honeyeater, White-fronted Melidectes)

606. **Melidectes belfordi**
Belford's Melidectes

Melirrhophetes belfordi De Vis, 1890, Ann. Rept. British New Guinea, 1888-1889: 60.

This and the next species hybridize and intergrade throughout much of eastern New Guinea. Designation of hybrid individuals is somewhat arbitrary. In considering *belfordi* and *rufocrissalis* as distinct species we follow the treatment of Gilliard (1959). More study of variation in this group is needed.

(Belford's Honeyeater)

607. **Melidectes rufocrissalis** Yellow-browed Melidectes

Melirrhophetes rufocrissalis Reichenow, 1915, Journ. f. Ornith. 63: 126.

(Reichenow's Melidectes, Yellow-browed Honeyeater)

608. **Melidectes foersteri** Huon Melidectes

Melirrhophetes foersteri Rothschild and Hartert, 1911, Bull. Brit. Ornith. Cl. 29: 12.

(Foerster's Honeyeater, Foerster's Melidectes)

609. **Melidectes torquatus** Ornate Melidectes

Melidectes torquatus Sclater, 1873, Proc. Zool. Soc. London: 694, pl. 55.

(Cinnamon-breasted Wattle Bird, Ornate Honeyeater)

610. **Melipotes gymnops** Western Smoky Honeyeater

Melipotes gymnops Sclater, 1873, Proc. Zool. Soc. London: 695, pl. 56.

(Spot-bellied Honeyeater, Arfak Melipotes)

611. **Melipotes fumigatus** Common Smoky Honeyeater

Melipotes fumigatus A. B. Meyer, 1886, Zeitschr. Ges. Ornith. 3: 22, pl. 4, fig. 1.

(Smoky Honeyeater, Common Melipotes)

612. **Melipotes ater** Spangled Honeyeater

Melipotes ater Rothschild and Hartert, 1911, Bull. Brit. Ornith. Cl. 29: 13.

(Huon Melipotes)

613. **Ramsayornis modestus** Brown-backed Honeyeater

Glyciphila modesta G. R. Gray, 1858, Proc. Zool. Soc. London: 174.

(Modest Honeyeater)

614. **Conopophila albogularis** Rufous-banded Honeyeater

Entomophila ? albogularis Gould, 1842 (1843), Proc. Zool. Soc. London: 137.

(Rufous-breasted Honeyeater)

FAMILY **ESTRILDIDAE**: MANNIKINS, PARROT-FINCHES AND ALLIES

Treatment follows Mayr (1968b). For familial relationships see Bentz (1979). Note: There are sight records of the House Sparrow *Passer domesticus* from Port Moresby (Ashford 1978). Were these individuals artificially assisted in passage by ships travelling from Australia to New Guinea?

615. **Erythrura trichroa** Blue-faced Parrot-Finch

Fringilla trichroa Kittlitz, 1835, Mem. Prés. Acad. Imp. Sci. St. Pétersbourg 2: 8, tab. 10.

(Blue-faced Parrot-Mannikin, Blue-faced Finch)

616. **Erythrura papuana** Papuan Parrot-Finch

Erythrura trichroa papuana Hartert, 1900, Novit. Zool. 7: 7.

(Large-billed Parrotfinch, Papuan Parrot-Mannikin)

617. **Lonchura leucosticta** White-spotted Mannikin

Munia leucosticta D'Albertis and Salvadori, 1879, Ann. Mus. Civ. Genova 14: 88.

618. **Lonchura tristissima** Streak-headed Mannikin

Munia tristissima Wallace, 1865, Proc. Zool. Soc. London: 479.

Early reports of *L. punctulata* near Port Moresby (Wheeler 1967) were perhaps misidentifications of *tristissima*. More recent reports may be of escaped cage birds.

(Streaked Mannikin)

619. **Lonchura vana** Grey-banded Mannikin

Munia vana Hartert, 1930, Novit. Zool. 36: 42.

(Arfak Mannikin)

620. **Lonchura caniceps** Grey-headed Mannikin

Munia caniceps Salvadori, 1876, Ann. Mus. Civ. Stor. nat. Genova 9: 38.

621. **Lonchura nevermanni** Grey-crowned Mannikin

Lonchura nevermanni Stresemann, 1934, Ornith. Monatsb. 42: 101.

(White-crowned Mannikin)

622. **Lonchura grandis** Grand Mannikin

Munia grandis Sharpe, 1882, Journ. Linn. Soc. London, Zool. 16: 319.

(Great-billed Mannikin)

623. **Lonchura stygia** Black Mannikin

Lonchura stygia Stresemann, 1934, Ornith. Monatsb. 42: 102.

624. **Lonchura spectabilis** Hooded Mannikin

Donacola spectabilis Sclater, 1879, Proc. Zool. Soc. London: 449, pl. 37, fig. 2.

(New Britain Mannikin)

625. **Lonchura castaneothorax** Chestnut-breasted Mannikin

Amadina castaneothorax Gould, 1837, Synops. Bds. Austr. 2: pl. 21, fig. 2.

((Chestnut-breasted Finch)

626. **Lonchura teerinki** Black-breasted Mannikin

Lonchura teerinki Rand, 1940, Am. Mus. Novit. 1072: 14.

(*Lonchura terrinki* [sic], Grand Valley Mannikin)

627. **Lonchura montana** Western Alpine Mannikin

Lonchura montana Junge, 1939, Nova Guinea, new ser., 3: 67.

(Snow Mountain Mannikin)

628. **Lonchura monticola** Eastern Alpine Mannikin

Munia monticola De Vis, 1897, Ibis: 387.

(Alpine Mannikin)

629. **Oreostruthus fuliginosus** Mountain Firetail

Oreospiza fuliginosa De Vis, 1897, Ibis: 389.

(Crimson-sided Mountain Finch, Alpine Finch)

630. **Neochmia phaeton** Crimson Finch

Fringilla phaeton Hombron and Jacquinot, 1841, Ann. Sci. Nat., Paris, ser. 2, 16: 314.

(*Neochmia evangelinae*)

FAMILY **STURNIDAE**: STARLINGS AND MYNAS

Treatment follows Amadon (1962).

631. **Aplonis cantoroides** Singing Starling

Calornis cantoroides G. R. Gray, 1861 (1862), Proc. Zool. Soc. London: 431.

632. **Aplonis mysolensis** Moluccan Starling

Calornis mysolensis G. R. Gray, 1861 (1862), Proc. Zool. Soc. London: 431.

633. **Aplonis magna** Long-tailed Starling

Lamprotornis magnus 'von Rosenberg' Schlegel, 1871 (1873), Nederl. Tijdschr. Dierk. 4: 18.

634. **Aplonis metallica** Metallic Starling

Lamprotornis metallicus Temminck, 1824, Pl. Col., livr. 45: pl. 266.

(Shining or Colonial Starling)

635. **Aplonis mystacea** Yellow-eyed Starling

Calornis mystacea Ogilvie-Grant, 1911, Bull. Brit. Ornith. Cl. 29: 28.

(Grant's or Mimic Starling)

636. **Sturnus vulgaris** Common Starling

Sturnus vulgaris Linnaeus, 1758, Syst. Nat., ed. 10, 1: 167.

Recorded in region by Coates (1970).

(Starling)

637. **Mino anais** Golden Myna

Sericulus Anais Lesson, 1839, Rev. Zool.: 44.

(*Melanopyrrhus anais*, Golden Grackle, Golden-breasted Myna)

638. **Mino dumontii** Yellow-faced Myna

Mino dumontii Lesson, 1827, Bull. Sci. Nat. Ferussac 10: 159.

(*Mino dumonti*, Orange-faced Grackle)

FAMILY **ORIOLIDAE:** ORIOLES

Treatment follows Greenway (1962), with emendation.

639. **Oriolus szalayi** Brown Oriole

Mimeta szalayi Madarász, 1900, Termés. Füzet. 24: 76, 80.

(New Guinea Oriole)

640. **Oriolus sagittatus** Olive-backed Oriole

Coracias sagittata Latham, 1801, Index Ornith., Suppl.: 26.

(White-bellied Oriole)

641. **Oriolus flavocinctus** Yellow Oriole

Mimetes flavocinctus Vigors, in King, 1826 (1827), Surv. Intertrop. Coasts Austr. 2: 419.

(Green Oriole)

642. **Sphecotheres viridis** Figbird

Sphecotera [sic] viridis Vieillot, 1816, Analyse: 68.

Treatment follows Ford (1975).

(*Sphecotheres vieilloti*, Green Figbird)

FAMILY **DICRURIDAE:** DRONGOS

Treatment follows Vaurie (1962).

643. **Chaetorhynchus papuensis** Mountain Drongo

Chaetorhynchus papuensis A. B. Meyer, 1874, Sitzungsb. Akad. Wiss. Wien 69 (1): 493.

(Pygmy Drongo)

644. **Dicrurus hottentottus** Spangled Drongo

Corvus hottentottus Linnaeus, 1766, Syst. Nat., ed. 12, 1: 155.

White & Bruce (1985) split the widespread '*hottentottus*' of Vaurie (1962) into several species, and consider the Papuan form to be distinct (*D. bracteatus*). Certainly more study is needed of this Indo-Australian complex, which we agree probably contains more than a single species.

(*Chibia bracteata*, Hair-crested Drongo)

FAMILY **GRALLINIDAE:** MAGPIE-LARKS

Treatment follows Mayr (1962).

645. **Grallina cyanoleuca** Australian Magpie-lark

C[orvus] cyanoleucus Latham, 1801, Index Ornith., Suppl.: 25.

First recorded from New Guinea by Lindgren (1971). M. LeCroy (*in litt.*) notes, however, that there is a specimen in New York (AMNH) labelled as taken on the Laloki River (near Port Moresby) by the Dentons in 1883.

(Mudlark, Magpie Lark, Australian Magpie-lark, Peewit)

646. **Grallina bruijni** Torrent-lark

Grallina bruijni Salvadori, 1875, Ann. Mus. Civ. Stor. nat. Genova 7: 929.

(*Promareopsis bruijni*)

FAMILY **ARTAMIDAE:** WOOD-SWALLOWS

Treatment follows Mayr (1962).

647. **Artamus leucorhynchus** White-breasted Wood-swallow

Lanius leucorhyn[chus] Linnaeus, 1771, Mantissa Plant.: 524.

(Lesser or Whiterumped Wood Swallow)

648. **Artamus maximus** Great Wood-swallow

Artamus maximus A. B. Meyer, 1874, Sitzungsb. Akad. Wiss. Wien 69: 203.

(Greater or Black-breasted Woodswallow)

649. **Artamus cinereus** Black-faced Wood-swallow

Artamus cinereus Vieillot, 1817, Nouv. Dict. Hist. Nat., nouv. éd. 17: 297.

FAMILY **CRACTICIDAE**: BUTCHERBIRDS AND ALLIES

Treatment follows Amadon (1962) and Sibley & Ahlquish (1984).

650. **Cracticus cassicus** Hooded Butcherbird
Rhamphastos cassicus Boddaert, 1783, Table Pl. Enlum.: 38.
(Black-headed Butcherbird)

651. **Cracticus louisiadensis**
Tagula Butcherbird
Cracticus louisiadensis Tristram, 1889, Ibis: 555.
(Louisiade Butcherbird)

652. **Cracticus mentalis**
Black-backed Butcherbird
Cracticus mentalis Salvadori and D'Albertis, 1875 (1876), Ann. Mus. Civ. Stor. nat. Genova 7: 824.
(White-throated Butcherbird)

653. **Cracticus quoyi** Black Butcherbird
Barita Quoyi Lesson, 1827, Bull. Sci. Nat. Férussac 10: 289.

654. **Gymnorhina tibicen** Australian Magpie
C[oracias] tibicen Latham, 1801, Index Ornith., Suppl.: 27.
(Black-backed Magpie)

655. **Peltops blainvillii** Lowland Peltops
Eurylaimus Blainvillii Lesson and Garnot, 1827, Voy. Coquille, Atlas: pl. 19, fig. 2.
The genus *Peltops*, long considered an aberrant flycatcher, is now considered to be most closely related to the Cracticidae by Sibley & Ahlquist (1984).
(Clicking Shieldbill, Lowland Peltops Flycatcher)

656. **Peltops montanus** Mountain Peltops
Peltops montanus Stresemann, 1921, Anz. Ornith. Ges. Bayern. 1: 35.
(Singing Shieldbill, Mountain Peltops Flycatcher)

FAMILY **PTILONORHYNCHIDAE**: BOWERBIRDS

Treatment follows Mayr (1962), with emendation. Schodde (1976) proposed unifying the bowerbirds and birds of paradise into a single family; however, we follow Stonor (1937), Sibley & Ahlquist (in press) and others in maintaining two discrete families.

657. **Ailuroedus buccoides**
White-eared Catbird
Kitta buccoides Temminck, 1835, Pl. Col., livr. 97: pl. 575.
(White-throated Catbird)

658. **Ailuroedus melanotis** Spotted Catbird
Ptilonorhynchus melanotis G. R. Gray, 1858, Proc. Zool. Soc. London: 181.
Treatment follows Gilliard (1969). Some authorities (e.g. Mayr & Jennings 1952) considered this taxon to be conspecific with *Ailuroedus crassirostris*.
(Black-eared Catbird)

659. **Archboldia papuensis**
Archbold's Bowerbird
Archboldia papuensis Rand, 1940, Am. Mus. Novit. 1072: 9.
The population *sanfordi*, from the region between Tari and Mount Hagen, has been considered by some authorities to be a full species (*cf.* Peckover & Filewood 1976).
(Tomba Bowerbird, Sanford's Bowerbird)

660. **Amblyornis inornatus**
Vogelkop Bowerbird
Ptilorhynchus inornatus Schlegel, 1871 (1873), Nederl. Tijdsch. Dierk. 4: 51.
(Gardener or Plain Bowerbird, Brown Gardenerbird, Vogelkop Gardener Bower Bird)

661. **Amblyornis macgregoriae**
Macgregor's Bowerbird
Amblyornis macgregoriae De Vis, 1890, Ann. Rept. Brit. New Guinea, 1888-1889: 61.
(Macgregor's Gardenerbird, Macgregor's Bower Bird)

662. **Amblyornis subalaris**
Streaked Bowerbird

Amblyornis subalaris Sharpe, 1884, Journ. Linn. Soc. London, Zool. 17: 408.

(Striped Bower Bird, Striped Gardener Bowerbird, Eastern Gardenerbird)

663. **Amblyornis flavifrons** Golden-fronted Bowerbird

Amblyornis flavifrons Rothschild, 1895, Novit. Zool. 2: 480, (fig. Novit. Zool. 3: pl. 1, fig, 3 and 4.)

(Yellow-fronted Gardenerbird, Yellow-fronted Gardener Bowerbird, Yellow-fronted Bower Bird)

664. **Sericulus aureus** Flame Bowerbird

Coracias aurea Linnaeus, 1758, Syst. Nat., ed. 10, 1: 108.

(*Xanthomelus aureus*, Golden Bowerbird, Golden Regent Bower Bird)

665. **Sericulus bakeri** Fire-maned Bowerbird

Xanthomelus bakeri Chapin, 1929, Am. Mus. Novit. 367: 1.

(Adelbert Bowerbird, Beck's Bowerbird, Adelbert Regent Bowerbird)

666. **Chlamydera lauterbachi** Yellow-breasted Bowerbird

Chlamydodera lauterbachi Reichenow, 1897, Ornith. Monatsb. 5: 24.

(Lauterbach's or Yellow-bellied Bowerbird)

667. **Chlamydera cerviniventris** Fawn-breasted Bowerbird

Chlamydera cerviniventris Gould, 1850, in Jardine's Contr. Ornith.: 106. (also Proc. Zool. Soc. London, 1850 (1851): 201).

FAMILY **PARADISAEIDAE:** BIRDS OF PARADISE

Treatment follows the sequence and nomenclature of Gilliard (1969) with emendations. For the subfamily Cnemophilinae we follow the treatment of Diamond (1972); additionally, we follow Diamond in (1) placement of *Paradigalla* near *Manucodia*, (2) in merging *Phonygammus* with *Manucodia*, (3) in the sequence of species in *Manucodia*, (4) in placing *Drepanornis* into *Epimachus*. We do not advocate Diamond's enlarged *Lophorina* and instead follow Schodde (1976) in placing *Cicinnurus* and *Diphyllodes* in one genus separate from *Lophorina*. Spelling of the family name follows LeCroy (1983), *contra* McAlpine (1979).

668. **Cnemophilus macgregorii**
Crested Bird of Paradise

Cnemophilus macgregorii De Vis, 1890, Ann. Rept. Brit. New Guinea, 1888-1889: 62.

Considered to be a bowerbird by Mayr (1941).

(Sickle-crested or Multi-crested Bird of Paradise)

669. **Cnemophilus loriae** Loria's Bird of Paradise

Loria loriae Salvadori, 1894, Ann. Mus. Civ. Stor. nat. Genova, ser. 2, 14: 151.

Generic treatment follows Diamond (1972).

670. **Loboparadisea sericea** Yellow-breasted Bird of Paradise

Loboparadisea sericea Rothschild, 1896, Bull. Brit. Ornith. Cl. 6: 16.

(Wattle-billed Bird of Paradise)

671. **Macgregoria pulchra** Macgregor's Bird of Paradise

Macgregoria pulchra De Vis, 1897, Ibis: 251, pl. 7.

672. **Manucodia atra**
Glossy-mantled Manucode

Phonygama ater Lesson, 1830, Voy. Coquille, Zool. 1: 638.

We follow Filewood & Peckover (1978), in treating the genus *Manucodia* as feminine. Additional examination of the nomenclatural question supports their contention. According to the International Rules of Zoological Nomenclature, *Manucodia* should be feminine for three reasons: (1) the original author, Boddaert clearly intended the name to be feminine (see Art. 30 (b)(ii)); (2) use of the suffix *-ia* can only be considered feminine in classical usage (Art. 30 (a)(ii) examples) and as such, being applied to a name based on old Javanese, must be treated as feminine. (3) The source of the name '*manuk dewa*' means 'bird of the gods', and *dewa* refers to a male god (*devi* is the female); nonetheless, the old Javanese language has no gender for nouns (P. A. Taylor, pers. comm.), thus obviating Mayr's assertion (1962: 184) that the genus should be treated as masculine because it is based on a masculine root. The Rules state that when a name is based on neither a classical nor modern Indo-European word (or words) then the author's intent should be followed (Art. 30 (b)(ii)).

(*Manucodia ater*)

673. **Manucodia chalybata** Crinkle-collared Manucode

Paradisea Chalybata Pennant, 1781, Spec. Faun. Ind., in Forster, Ind. Zool.: 40 (based on d'Aubenton, Pl. Enlum.: 634).

Manucodia chalybatus, Green-breasted Manucode)

674. **Manucodia comrii** Curl-crested Manucode

Manucodia comrii Sclater, 1876, Proc. Zool. Soc. London: 459, pl. 42.

(Curl-breasted Manucode)

675. **Manucodia jobiensis** Jobi Manucode

Manucodia jobiensis Salvadori, 1875 (1876), Ann. Mus. Civ. Stor. nat. Genova 7: 969.

676. **Manucodia keraudrenii** Trumpet Manucode

Barita Keraudrenii Lesson and Garnot, 1826, Bull. Sci. Nat. Férussac 8: 110.

(*Phonygammus keraudrenii*, Trumpet Bird, Manucode)

677. **Paradigalla carunculata** Long-tailed Paradigalla

Paradigalla carunculata Lesson, 1835, Hist. Nat. Ois. Parad.: 242.

Sometimes has been treated as conspecific with *brevicauda* (Mayr 1941).

678. **Paradigalla brevicauda** Short-tailed Paradigalla

Paradigalla brevicauda Rothschild and Hartert, 1911, Novit. Zool. 18: 159.

679. **Ptiloris magnificus** Magnificent Riflebird

Falcinellus magnificus Vieillot, 1819, Nouv. Dict. Hist. Nat., nouv. éd. 28: 167.

(*Craspedophora magnifica, Lophorina magnifica*)

680. **Seleucidis melanoleuca** Twelve-wired Bird of Paradise

Paradisea melanoleuca Daudin, 1800, Traité Ornith. 2: 278.

(*Lophorina melanoleuca, Seleucidis ignota, Seleucides melanoleuca* or *ignotus*)

681. **Epimachus albertisi** Buff-tailed Sicklebill

Drepanephorus albertisi Sclater, 1873, Nature 8: 151 and Proc. Zool. Soc. London: 558, pl. 47.

Spelling of the specific epithet in all recent treatments has been *albertisii*. We re-examined both of Sclater's descriptions and found the spelling to be *albertisi*, which remains correct.

(*Drepanornis albertisii*, Black-billed Sicklebill, Black-billed Sicklebill Bird of Paradise, Short-tailed Sicklebill).

682. **Epimachus bruijnii** Pale-billed Sicklebill

Drepanornis Bruijnii Oustalet, 1880, Ann. Sci. Nat., ser. 6: 9: art. 5, p. 1; and Bull. Assoc. Sci. France (1880): 172.

(White-billed Sicklebill Bird of Paradise, Lowland Sicklebill).

683. **Epimachus fastuosus** Black Sicklebill

Promerops fastuosus Hermann, 1783, Tab. Aff.

Anim.: 194 (based on Pl. Enlum. pls. 638-639).

(*Epimachus fastosus* [sic], Black Sickle-billed Bird of Paradise)

684. **Epimachus meyeri** Brown Sicklebill

Epimachus meyeri Finsch, 1885, Zeitschr. Ges. Ornith. 2: 380.

(Brown Sickle-billed Bird of Paradise)

685. **Astrapia nigra** Arfak Astrapia

Paradisea nigra Gmelin, 1788, Syst. Nat. 401.

(Black Astrapia, Arfak Astrapia Bird of Paradise, Arfak Bird of Paradise)

686. **Astrapia splendidissima** Splendid Astrapia

Astrapia splendidissima Rothschild, 1895, Novit. Zool. 2: 59, pl. 5.

(Splendid Astrapia Bird of Paradise, Splendid Bird of Paradise)

687. **Astrapia mayeri** Ribbon-tailed Astrapia

Astrapia mayeri Stonor, 1939 (Febr.), Bull. Brit. Ornith. Cl. 59: 57.

(Ribbontail Astrapia, Ribbon-tailed Bird of Paradise)

688. **Astrapia stephaniae** Stephanie's Astrapia

Astrarchia stephaniae Finsch and Meyer, 1885, Zeitschr. Ges. Ornith. 2: 378.

(Princess Stephania Astrapia, Stephanie Astrapia, or Princess Stephanie Bird of Paradise)

689. **Astrapia rothschildi** Huon Astrapia

Astrapia rothschildi Foerster, 1906, Two New Bds. of Parad.: 2.

(Rothschild's Astrapia, Huon Bird of Paradise, Huon Astrapia Bird of Paradise)

690. **Lophorina superba** Superb Bird of Paradise

Paradisea Superba Pennant, 1781, Spec. Faun. Ind., in Forster, Ind. Zool.: 40, (based on d'Aubenton, Pl. Enlum: pl. 362.)

691. **Parotia wahnesi** Wahnes' Parotia

Rothschild, 1906, Two New Bds. of Parad.: 2

(Wahnes's Six-wired Parotia, Huon Parotia, Wahnes' Six-wired Bird of Paradise)

692. **Parotia sefilata** Western Parotia

Paradisea sefilata Pennant, 1781, Spec. Fauna Ind., in Forster, Ind. Zool: 40.

(Arfak Parotia, Arfak Six-wired Parotia, Arfak Six-wired Bird of Paradise)

693. **Parotia carolae** Carola's Parotia

Parotia carolae A. B. Meyer, 1894, Bull. Brit. Ornith. Cl. 4: 6.

(Carola Parotia, Queen Carola's Parotia, Queen Carola's Six-wired Bird of Paradise)

694. **Parotia lawesii** Lawes' Parotia

Parotia lawesii Ramsay, 1885, Proc. Linn. Soc. New South Wales 10: 247.

Schodde & McKean (1973) treat the eastern population *helenae* as a full species; however, we follow Gilliard (1969) and Diamond (in press) in keeping *helenae* as a race within *lawesii*.

(*Parotia lawesi*, Lawes's Six-wired Parotia, includes Helena's Parotia)

695. **Pteridophora alberti** King of Saxony Bird of Paradise

Pteridophora alberti A. B. Meyer, 1894, Bull. Brit. Ornith. Cl. 4: 11.

(Kissaba)

696. **Cicinnurus regius** King Bird of Paradise

Paradisaea regia Linnaeus, 1758, Syst. Nat., ed. 10, 1: 110.

(*Lophorina regia*)

697. **Cicinnurus magnificus** Magnificent Bird of Paradise

Paradisea Magnifica Pennant, 1781, Spec. Faun. Ind., in Forster, Ind. Zool.: 40, (based on d'Aubenton, Pl. Enlum.: pl. 631).

(*Diphyllodes magnificus*, *Lophorina magnifica*)

698. **Cicinnurus respublica**
Wilson's Bird of Paradise

Loph[orina] respublica Bonaparte, 1850 (Febr.), Compt. Rend. Acad. Sci., Paris 30: 131.

(*Diphyllodes respublica*, Waigeu or Waigeo Bird of Paradise)

699. **Paradisaea raggiana**
Raggiana Bird of Paradise

Paradisea raggiana Sclater, 1873, Proc. Zool. Soc. London: 559.

Formerly considered a subspecies of *P. apoda* (Mayr 1941).

(Count Raggi's Bird of Paradise)

700. **Paradisaea apoda**
Greater Bird of Paradise

Paradisaea apoda Linnaeus, 1758, Syst. Nat., ed. 10, 1: 110.

701. **Paradisaea minor**
Lesser Bird of Paradise

Paradisea minor Shaw, 1809, Gen. Zool 7(2): 486.

702. **Paradisaea decora**
Goldie's Bird of Paradise

Paradisea decora Salvin and Godman, 1883, Ibis: 131.

703. **Paradisaea rubra** Red Bird of Paradise

Paradisaea rubra Daudin, 1800, Traité Ornith. 2: 271.

704. **Paradisaea guilielmi**
Emperor Bird of Paradise

Paradisea Guilielmi Cabanis, 1888, Journ. f. Ornith. 36: 119.

(Emperor of Germany's Bird of Paradise)

705. **Paradisaea rudolphi**
Blue Bird of Paradise

Paradisornis Rudolphi Finsch, 1885, Zeitschr. Ges. Ornith. 2: 385, pl. 20.

FAMILY **CORVIDAE:** CROWS

Treatment follows Blake & Vaurie (1962).

706. **Corvus fuscicapillus**
Brown-headed Crow

Corvus fuscicapillus Gray, 1859, Proc. Zool. Soc. London: 157.

707. **Corvus tristis** Grey Crow

Corvus tristis Lesson and Garnot, 1827, Bull. Sci. Nat. Férussac 10: 291.

(*Gymnocorvus tristis*, Bare-eyed Crow)

708. **Corvus orru** Torresian Crow

C[orvus] orru Bonaparte, 1850 (1851), Consp. Gen. Av. 1: 385.

(Australian Crow, Papuan Crow)

REFERENCES

AMADON, D. 1959. Remarks on the subspecies of the Grass Owl, *Tyto capensis*. *J. Bombay Nat. Hist. Soc.* 56: 344-346.

——— 1962. Cracticidae; Sturnidae. Pp. 75-120; 166-171; *in* E. Mayr and J. C. Greenway (eds.), Checklist of Birds of the World, Volume 15. Cambridge: Museum comp. Zool.

——— 1978. Remarks on the taxonomy of some Australasian raptors. *Emu* 78: 115-118.

ASHFORD, R. W. 1978. First record of House Sparrows for Papua New Guinea. *Emu* 78: 36.

BAKER, A. 1975. Systematics and evolution of Australian oystercatchers. *Emu* 74: 277.

BAKER-GABB, D. 1979. Remarks on the taxonomy of the Australasian Harrier (*Circus approximans*). *Notornis* 28: 325-329.

BEEHLER, B. 1978. The status of *Sericornis nigroviridis*. *Condor* 80: 115-116.

——— 1980. Black Swan *Cygnus atratus*: A new species for the New Guinea region. *PNG Bird Soc. Newsl.* 173-174: 3.

———, T. K. PRATT, & D. A. ZIMMERMAN, 1986. Birds of New Guinea. Princeton: Princeton Univ. Press.

BELL, H. L. 1966a. Some wader observations from New Guinea. *Emu* 66: 32.

——— 1966b. Observations. [First record of *Larus novaehollandiae* from New Guinea]. *NG Bird Soc. Newsl.* 5: 1.

——— 1968. The Fairy Martin, a new bird for New Guinea. *Emu* 68: 5.

——— 1970. Field notes on the birds of Amazon Bay, Papua. *Emu* 70: 23-26.

——— 1979. Possible occurrence of Sand Martin at Finschhafen. *PNG Bird Soc. Newsl.* 155: 12.

——— 1982. Social organization and feeding of the Rufous Babbler *Pomatostomus isidori*. *Emu* 82: 7-11.

BENTZ, G. D. 1979. The appendicular myology and phylogenetic relationships of the Ploceidae and Estrildidae (Aves: Passeriformes). *Bull. Carnegie Mus. Nat. Hist.* 15: 1-25.

BERGGY, J. 1978. Bird observations in the Madang Province. *PNG Bird Soc. Newsl.* 148: 9-20.

BLAKE, E. R., & C. VAURIE. 1962. Corvidae. Pp. 204-284, *in* J. L. Peters (ed.), Check-list of Birds of the World, Volume 15. Cambridge: Mus. comp. Zool.

BOCK, W. J. 1958. A revision of the Ardeidae. *Am. Mus. Novit.* 1779: 1-49.

BOLES, W. E. 1979. The relationships of the Australo-Papuan flycatchers. *Emu* 79: 107-110.

——— 1981. The subfamily name of the monarch flycatchers. *Emu* 81: 50.

———, & N. W. LONGMORE. 1983. A new subspecies of treecreeper in the *Climacteris leucophaea* superspecies. *Emu* 83: 272-275.

BRODKORB, P. 1963. Catalogue of fossil birds. *Bull. Fla. State Mus.* 7: 179-293.

BROOKE, R. K. 1970. Taxonomic and evolutionary notes on the subfamilies, tribes, genera, and subgenera of the swifts (Aves: Apodidae). *Durban Mus. Novit.* 9: 13-24.

———, & P. A. CLANCY. 1981. The authorship of the generic and specific name of the Bat Hawk. *Bull. Brit. Orn. Cl.* 101: 371-372.

BROWN, L., & D. AMADON. 1968. Eagles, Hawks and Falcons of the World, 2 volumes. Feltham, Middlesex: Country Life Books.

COATES, B. J. 1970a. Observations. [First record of *Puffinus carneipes*]. *NG Bird Soc. Newsl.* 51: 1.

——— 1970b. The Common Starling—a bird new to New Guinea. *NG Bird Soc. Newsl.* 59: 3.

——— 1972. A Redshank near Port Moresby—a new bird for New Guinea. *NG Bird Soc. Newsl.* 81: 3.

——— 1973. The Pectoral Sandpiper—a new bird for Papua New Guinea. *NG Bird Soc. Newsl.* 92: 3-4.

COLLINS, R. T., & R. K. BROOKE. 1976. A review of the swifts of the genus *Hirundapus* (Aves: Apodidae). *Contrib. Sci. L.A. County Mus.* 282: 1-21.

CONDON, H. T. 1975. Checklist of the Birds of Australia, Part 1. Non-Passerines. Melbourne: RAOU.

CONNERS, P. G. 1983. Taxonomy, distribution, and evolution of the Golden Plovers (*Pluvialis dominica* and *P. fulva*). *Auk 100*: 607-620.

COOPER, W. T., & J. M. FORSHAW. 1977. The Birds of Paradise and Bowerbirds. Sydney: Collins.

CRAMP, S. (ed.). 1983. Handbook of the Birds of Europe, the Middle East, and North Africa, Volume 3. Oxford: Oxford Univ. Press.

CROME, F. H. J., & G. W. SWAINSON. 1974. Sight record of the Pied Harrier in northern New Guinea. *Emu* 74: 103.

DEIGNAN, H. G. 1964. Orthonychinae. Pp. 228-240, *in* E. Mayr, & R. A. Paynter, Jr. (eds.), Checklist of Birds of the World, Volume 10. Cambridge: Mus. comp. Zool.

DIAMOND, J. M. 1969. Preliminary results of an ornithological exploration of the North Coastal Range, New Guinea. *Am. Mus. Novit.* 2362: 1-67.

——— 1972. Avifauna of the Eastern Highlands of New Guinea. Publ. Nuttall Orn. Club No. 12.

——— In press. New distributional records and taxa from the outlying mountain ranges of Irian Jaya. *Emu.*

———, & M. LeCROY. In prep. The status of *Malurus campbelli*.

DORST, J., & J.-L. MOUGIN. 1979. Pelecaniformes. Pp. 155-193, *in* E. Mayr and G. W. Cottrell (eds.), Checklist of Birds of the World, Volume 1 (2nd ed.). Cambridge: Mus. comp. Zool.

DRAFFAN, R. D. W., S. T. GARNETT, & G. J. MALONE. 1983. Birds of the Torres Strait: an annotated list and biogeographic analysis. *Emu* 83: 207-234.

FILEWOOD, L. W. C. 1974. A new bird for PNG. *NG Bird Soc. Newsl.* 104: 3-4.

———, & W. S. PECKOVER. 1978. Scientific names used in Birds of New Guinea and Tropical Australia and the Handbook of New Guinea Birds. Wildlife in Papua New Guinea 78/12.

FINCH, B. W. 1979a. Observations [First record of *Anas acuta* for New Guinea]. *PNG Bird Soc. Newsl.* 151: 8.

——— 1979b. The Palearctic wader presence of the Port Moresby area—1979. *PNG Bird Soc. Newsl.* 162: 4-13.

——— 1980a. First record of Little Stint *Calidris minuta* for Papua New Guinea. *PNG Bird Soc. Newsl.* 163-164: 10-15.

——— 1980b. First PNG record of the Great Reed Warbler *Acrocephalus arundinaceus orientalis* at Aroa Lagoon, Central Province. *PNG Bird Soc. Newsl.* 163-164: 16-20.

——— 1980c. Welcome Swallow *Hirundo neoxena* at Aroa lagoon, Central Province, a new species for the New Guinea region. *PNG Bird Soc. Newsl.* 173-4: 4-5.

——— 1980d. Discovery of the Zitting Cisticola *Cisticola juncidis* in the Bensbach floodplain (Bulla-Balamuk) Western Province. *PNG Bird. Soc. Newsl.* 173-174: 13-17.

——— 1981a. Sight record of a possible Grey Falcon *Falco hypoleucus* on the Aroa Plains (Central Province). *PNG Bird Soc. Newsl.* 175-176: 11-13.

——— 1981b. Boat trip from Madang to Karkar. *PNG Bird. Soc. Newsl.* 181-182: 21.

——— 1982a. Sight record of the Obscure Berrypecker *Melanocharis arfakianus* [sic] near Owers Corner. Central Province. *PNG Bird Soc. Newsl.* 187-188: 11-12.

——— 1982b. Gray's Grasshopper-Warbler *Locustella fasciolata. PNG Bird Soc. Newsl.* 197-198: 40.

——— 1983a. Little Stint *Calidris minuta* at Aroa Lagoon, Central Province. Second record for New Guinea region. *PNG Bird Soc. Newsl.* 201-202: 3-4.

——— 1983b. Should the Yellow-tinted Honeyeater be on the New Guinea List? *PNG Bird Soc. Newsl.* 201-202: 11-13.

——— 1983c. Siberian Rubythroat *Erithacus calliops* [sic]. A new species for the Australasian region. *PNG Bird Soc. Newsl.* 201-202: 31.

FORD, J. 1975. Systematics and hybridization of figbirds *Sphecotheres. Emu* 75: 163-171.

——— 1978. Intergradation between the Varied and Mangrove Honeyeaters. *Emu* 78: 71-74.

——— 1981a. Morphological and behavioural evolution in the *Gerygone fusca* complex. *Emu* 81: 57-81.

——— 1981b. Evolution, distribution and stage of speciation in the *Rhipidura fuliginosa* complex in Australia. *Emu* 81: 128-144.

——— 1981c. Hybridization and migration in Australian populations of the Little and Rufous-breasted Bronze-Cuckoos. *Emu* 81: 209-222.

——— 1982. Hybrid phenotypes in male Figbirds *Sphecotheres viridis* in Queensland. *Emu* 82: 126-130.

——— 1983. Speciation in the ground-thrush complex *Zoothera dauma* in Australia. *Emu* 83: 141-151.

FORSHAW, J. M. 1973. Parrots of the World. London: Lansdowne.

FRIEDMANN, H. 1968. The evolutionary history of the avian genus *Chrysococcyx. Bull. U.S. natn. Mus.* 165: 1-137.

FRITH, C. B. 1979. Ornithological literature of the Papuan Subregion. *Bull. Am. Mus. nat. Hist.* 164: 379-465.

FRY, C. H. 1969. The evolution and systematics of the bee-eaters (Meropidae). *Ibis* 111: 557-592.

——— 1980. The evolutionary biology of kingfishers (Alcedinidae). *Living Bird* 18: 113-160.

FULLER, ERROL. 1979. Hybridization amongst the Paradisaeidae. *Bull. Brit. Orn. Cl.* 99: 145-152.

GALBRAITH, I. C. J. 1967. The Black-tailed and Robust Whistlers *Pachycephala melanura* as a species distinct from the Golden Whistler *P. pectoralis. Emu* 66: 289-294.

GILLIARD, E. T. 1959. The ecology of hybridization in New Guinea honeyeaters (Aves). *Am. Mus. Novit.* 1937: 1-26.

——— 1969. The Birds of Paradise and Bower Birds. London: Weidenfeld & Nicolson.

———, & M. LeCROY. 1970. Notes on birds from the Tamrau Mountains, New Guinea. *Am. Mus. Novit.* 2420: 1-28.

GOODWIN, D. 1970. Pigeons and Doves of the World. London: British Mus. (Nat. Hist.).

GREENSMITH, A. 1975. Some notes on Melanesian seabirds. *Sunbird* 6: 77-89.

GREENWAY, J. C. 1962. Oriolidae. Pp. 122-137, *in* E. Mayr, & J. C. Greenway (eds.), Checklist of Birds of the World, Vol. 15. Cambridge: Mus. comp. Zool.

——— 1967. Climacteridae. Pp. 162-166, *in* R. A. Paynter, Jr. (ed.), Checklist of Birds of the World, Volume 12. Cambridge: Mus. comp. Zool.

HANCOCK, J. 1984. Aerial stretch display of the eastern race of the Great White Heron *Egretta alba. Ibis* 126: 92-94.

———, & H. ELLIOT. 1978. The Herons of the World. London: London Editions.

HARRISON, C. J. O., & PARKER, S. A. 1966. The taxonomic affinity of the New Guinea genera *Paramythia* and *Oreocharis. Bull. Brit. Orn. Cl.* 86: 15-20.

HERON, M. 1977a. First New Guinea record of the Asian Dowitcher. *NG Bird Soc. Newsl.* 129: 9.

——— 1977b. The Black-tailed Whistler *Pachycephala melanura* in mainland Papua New Guinea. *NG Bird. Soc. Newsl.* 131: 11-12.

———1978. Waders of the New Guinea region. *NG Bird Soc. Newsl.* 149-150: 8-15.

HOLYOAK, D. T. 1970. The relation of the parrot genus *Opopsitta* to *Psittaculirostris. Emu* 70: 198.

——— 1973. Comments on taxonomy and relationships in the parrot subfamilies Nestorinae, Loriinae and Platycercinae. *Emu* 73: 157-176.

HOOGERWERF, A. 1964. On birds new for New Guinea or with a larger range than previously known. *Bull. Brit. Orn. Cl.* 84: 70-77; 94-96; 118-124; 142-148; 153-161.

HUDSON, R., & J. W. H. CONROY. 1975. Southern Giant Petrel in New Guinea. *Emu* 75: 43.

IREDALE, T. 1956. Birds of New Guinea, 2 volumes. Melbourne: Georgian House.

JEHL, J. 1968. Relationships in the Charadrii (shorebirds): a taxonomic study based on color patterns of downy young. *Mem. San Diego Soc. Nat. Hist.* 3: 16-23.

JOHNSGARD, P. A. 1979. Anseriformes. Pp. 425-506, *in* E. Mayr, & G. W. Cottrell (eds.), Checklist of Birds of the World, Volume 1, 2nd ed.

JOHNSTONE, R. 1981. Notes on the distribution, ecology and taxonomy of the Red-crowned Pigeon (*Ptilinopus regina*) and Torres Strait Pigeon (*Ducula bicolor*) in Western Australia. *Rec. West. Austr. Mus.* 9: 7-22.

JOUANIN, C., & J.-L. MOUGIN. 1979. Procellariiformes. Pp. 48-121, *in* E. Mayr, & G. W. Cottrell (eds.), Checklist of Birds of the World, Volume 1, 2nd ed. Cambridge: Mus. comp. Zool.

JUNGE, G. C. A. 1953. Zoological results of the Dutch New Guinea expedition, 1939, no. 5. The birds. Zool. Verh. 20: 1-77.

KAHL, M. P. 1979. Ciconiiformes. Pp. 245-252, *in* E. Mayr, & G. W. Cottrell (eds.), Checklist of Birds of the World, Volume 1 (2nd ed.). Cambridge: Mus. comp. Zool.

KEAST, J. A. 1957. Variation and speciation in the genus *Climacteris* Temminck (Aves: Sittidae). Aust. J. Zool. 5: 474-495.

——— 1977. Relationships of the New Guinean Red-backed 'Warbler' *Eugerygone rubra. Emu* 77: 228-229.

KEMP, A. C. 1978. A review of the hornbills: biology and radiation. *Living Bird* 17: 105-136.

KING, B. 1979. New distributional records and field notes for some New Guinea birds. *Emu* 79: 146-148.

———, & E. C. DICKINSON. 1975. A Field Guide to the Birds of South-East Asia. Boston: Houghton Mifflin Co.

LECROY, M. 1969. *Acrocephalus arundinaceus orientalis,* first record in New Guinea. *Emu* 69: 119-120.

——— 1983. The spelling of *Semioptera wallacii* (Paradisaeidae). *Bull. Brit. Orn. Cl.* 103: 4.

LINDGREN, E. 1971. Records of new and uncommon species for the island of New Guinea. *Emu* 71: 134-136.

——— 1974. Another record of the Ruff *Philomachus pugnax* in Papua-New Guinea. *NG Bird Soc. Newsl.* 104: 4-5.

LONGMORE, N. W., & W. E. BOLES. 1983. Description and systematics of the Eungella Honeyeater *Meliphaga hindwoodi,* a new species of honeyeater from central eastern Queensland. *Emu* 83: 59-65.

MACDONALD, J. D. 1973. Birds of Australia. Sydney: Reed.
MACKAY, R. D. 1976. A new bird for PNG, Tawny Frogmouth *Podargus strigoides*. *PNG Bird Soc. Newsl.* 121: 10.
MARIEN, D. 1950. Notes on some Asiatic Meropidae (birds). *J. Bombay nat. Hist. Soc.* 49: 151-164.
MARSHALL, J. T. 1978. Systematics of smaller Asian night birds based on voice. *AOU Orn. Monogr.* No. 25: 1-58.
MASON, I. J. 1982. The identity of certain Australian types referred to the Cuculidae. *Bull. Brit. Orn. Cl.* 102: 99-106.
MATHEWS, G. M. 1927-1930. Systema Avium Australasianarum, two parts. London: BOU.
MAYR, E. 1938. Birds collected during the Whitney South Seas Expedition. XL. Notes on New Guinea birds. *Am. Mus. Novit.* 1006: 1-16.
——— 1941. List of New Guinea Birds. New York: Amer. Mus. Nat. Hist.
——— 1944. Birds collected during the Whitney South Seas Expedition No. 54. Notes on some genera from the Southwest Pacific. *Am. Mus. Novit.* 1269: 1-14.
——— 1962. Grallinidae; Artamidae; Ptilonorhynchidae; Paradisaeidae. Pp. 159-160; 160-165; 172-180; 181-203; *in* E. Mayr, & J. C. Greenway (eds.), Checklist of Birds of the World, Volume 15. Cambridge: Mus. comp. Zool.
——— 1967. Pachycephalinae. Pp. 3-51, *in* R. A. Paynter, Jr. (ed.), Checklist of Birds of the World, Volume 12. Cambridge: Mus. comp. Zool.
——— 1968a. *Larius* Boddaert, 1783 (Aves): proposed suppression under the plenary powers. Z. N. (S.) 1833. *Bull. zool. Nomencl.* 25: 52-54.
——— 1968b. Australo-Papuan Estrildidae. Pp. 306-397, *in* R. A. Paynter, Jr. (ed.), Checklist of Birds of the World, Volume 14. Cambridge: Mus. comp. Zool.
——— 1979a. Pittidae. Pp. 310-329. *In* M. A. Traylor (ed.), Checklist of Birds of the World, Volume 8. Cambridge: Mus. comp. Zool.
——— 1979b. Struthioniformes. Pp. 3-10, *in* E. Mayr, & G. W. Cottrell (eds.), Checklist of Birds of the World, Volume 1 (2nd ed.). Cambridge: Mus. comp. Zool.
——— 1985. Acanthizidae; Rhipiduridae; Monarchidae; Eopsaltriidae. *In* E. Mayr, & G. W. Cottrell (eds.), Checklist of Birds of the World, Volume 11. Cambridge: Mus. comp. Zool.
———, & E. T. GILLIARD. 1951. New species and subspecies of birds from the highlands of New Guinea. *Am. Mus. Novit.* 1524: 1-15.
———, & G. W. COTTRELL (eds.). 1979. Checklist of Birds of the World, Volume 1 (2nd ed.). Cambridge: Mus. comp. Zool.
———, & J. C. GREENWAY. 1956. Sequence of Passerine families (Aves). *Breviora* 58: 1-11.
———, & K. JENNINGS. 1952. Geographic variation and plumages in Australian bowerbirds. *Am. Mus. Novit.* 1602: 1-18.
———, & R. A. PAYNTER, JR. (eds.) 1964. Checklist of Birds of the World, Volume 10. Cambridge: Mus. comp. Zool.
———, & L. L. SHORT. 1970. Species taxa of North American Birds. Publ. Nuttall Orn. Club, No. 9.
MCALPINE, D. K. 1979. The correct name and authority for Wallace's Standard Wing (Passeriformes, Paradiseidae). *Bull. Brit. Orn. Cl.* 99: 108-110.
MCKEAN, J. L. 1969. The brush tongue of the Artamidae indicates more study is needed of relationships. *Bull. Brit. Orn. Cl.* 89: 129-130.
——— 1978. Some remarks on the taxonomy of Australasian oystercatchers. *Sunbird* 9: 3-6.
———, & VERNON. 1971. New records of tube-nosed birds (order Procellariiformes) from Queensland. *Mem. Qld. Mus.* 16: 141-144.
MEDWAY, L., & J. D. PYE. 1977. Echolocation and systematics of swiftlets. Pp. 225-238, *in* B. Stonehouse & C. M. Perrins (eds.), Evolutionary Ecology. Baltimore: University Park Press.
MEES, G. L. 1966. A new subspecies of *Anthreptes malaccensis* (Scopoli) from the Soela Islands (Aves, Nectariniidae). *Zool. Meded. (Leiden)* 41(18): 255-257.
——— 1969. A systematic revision of the Indo-Australian Zosteropidae, Part III. *Zool. Verhand (Leiden)* 102: 1-390.
——— 1979. Die Nachweise van *Cuculus canorus* im Indo-Australischen. *Mitt. Zool. Mus. Berl.* 55 (Suppl.): 127-134.
——— 1982a. [Review of] Nocturnal Birds of Australia, by R. Schodde and I. J. Mason. *Emu* 82: 182-184.
——— 1982b. Birds from the lowlands of southern New Guinea (Merauke and Koembe). *Zool. Ver. (Leiden)* 191: 1-188.
MILLER, A. H. 1964. A new species of warbler from New Guinea. *Auk* 81: 1-4.
MOCK, D. 1977. [Review of] Systematics and evolutionary relationships among the herons (Ardeidae), by R. B. Payne and C. J. Risley. *Bird-Banding* 48: 81-84.
MORDUE, T. A. 1981. Buff-breasted Sandpiper at Sangara, Oro Province. *PNG Bird Soc. Newsl.* 181-182: 34-35.
———, & B. W. FINCH. 1983. The complete species index to the PNG Bird Society Newsletter. Waigani: Office of Environment and Conserv. [Mimeo].
MORONY, J. J., W. J. BOCK, & J. FARRAND. 1975. Reference List of Birds of the World. New York: American Museum of Natural History.
OLSON, S. L. 1973. A classification of the Rallidae. *Wilson Bull.* 85: 381-416.
——— 1980. The significance of distribution of the Megapodiidae. *Emu* 80: 21-24.
OPIT, G. 1975. Observations. *PNG Bird Soc. Newsl.* 115: 5.
ORENSTEIN, R. I. 1979. Wing-flashing in *Eugerygone rubra*. *Emu* 79: 43-44.
PARKER, S. A. 1977. The distribution and occurrence in South Australia of owls of the genus *Tyto*. *South Austr. Orn.* 27: 207-215.
——— 1981. Prolegomenon to further studies in the *Chrysococcyx malayanus* group (Aves: Cuculidae). *Zool. Verh. (Leiden)* 187: 1-56.
——— 1982. The relationships of the Australo-Papuan treecreepers and sittellas. *South Austr. Orn.* 28: 197-200.
PAYNE, R. B. 1979. Ardeidae. Pp. 193-244, *in* E. Mayr, & G. W. Cottrell (eds.), Checklist of Birds of the World, Volume 1 (2nd ed.). Cambridge: Mus. comp. Zool.
PECKOVER, W. S., & L. W. C. FILEWOOD. 1976. Birds of New Guinea and Tropical Australia. Sydney: Reed.
PETERS, J. L. 1931-1945. Checklist of Birds of the World, Volumes 1-5. Cambridge: Harvard Univ. Press.
——— 1940. A genus for *Eupetes caerulescens* Temminck. *Auk* 57: 94.
——— 1960. Alaudidae; Hirundinidae; Campephagidae. Pp. 3-80; 80-130; 167-221; *in* E. Mayr, & J. C. Greenway (eds.), Checklist of Birds of the World, Volume 9. Cambridge: Mus. comp. Zool.
———, & E. MAYR. 1960. Campephagidae. *In* J. L. Peters (ed.), Checklist of Birds of the World, Volume 9. Cambridge: Harvard Univ. Press.
PHILLIPS, A. R. 1973. On the supposed genus *Petrochelidon*. *Bull. Brit. Orn. Cl.* 93: 20.
PIZZEY, G. 1980. A Field Guide to the Birds of Australia. Sydney: Collins.

RAND, A. L. 1936. *Meliphaga analoga* and its allies. *Am. Mus. Novit.* 872: 1-23.

——— 1941. New and interesting birds from New Guinea. *Am. Mus. Novit.* 1102: 1-15.

——— 1960. Laniidae. Pp. 309-365, *in* E. Mayr, & J. C. Greenway (eds.), Checklist of Birds of the World, Volume 9. Cambridge: Mus. comp. Zool.

——— 1967. Nectariniidae. Pp. 208-289, *in* J. L. Peters (ed.), Checklist of Birds of the World, Volume 12.

———, & E. T. GILLIARD. 1967. Handbook of New Guinea Birds. London: Weidenfeld and Nicolson.

READER'S DIGEST. 1976. Readers Digest Complete Book of Australian Birds. Sydney: Readers Digest Services.

RIPLEY, S. D. 1964a. Turdinae. Pp. 13-227. *In* E. Mayr, & R. A. Paynter, Jr. (eds.), Checklist of Birds of the World, Volume 10. Cambridge: Mus. comp. Zool.

——— 1964b. A systematic and ecological study of birds of New Guinea. *Yale Peabody Mus. Bull.* 19: 1-97.

——— 1977. Rails of the World. A Monograph of the Family Rallidae. Boston: David R. Godine.

——— 1982. A Synopsis of Birds of India and Pakistan. Bombay: Bomb. nat. Hist. Soc.

——— 1983. A record of Whitehead's Swiftlet *Collocalia whiteheadi* from Bougainville Island. *Bull. Brit. Orn. Cl.* 103: 82-84.

SALOMONSEN, F. 1967. Meliphagidae; Dicaeidae. Pp. 166-208; 338-450; *in* R. A. Paynter (ed.), Checklist of Birds of the World, Volume 12. Cambridge: Mus. comp. Zool.

——— 1983. Revision of the Melanesian swiftlets (Apodes, Aves) and their conspecific forms in the Indo-Australian and Polynesian region. *Kon. Danske Vidensk. Selskab Biolog. Skr.* 23: 5.

SANFT, K. (1960). Bucerotidae. Pp. 1-176, *in* R. Mertens, & W. Henning (eds.), Das Tierreich, Volume 76.

SCHODDE, R. 1975. Interim List of Australian Songbirds—Passerines. Melbourne: RAOU.

——— 1976a. Evolution in the birds-of-paradise and bowerbirds, a resynthesis. Proc. 16th Int. Orn. Congr.: 137-149.

——— 1976b. The mysterious origins of Australian birds. Pp. 602-605, *in* Reader's Digest Complete Book of Australian Birds. Sydney: Reader's Digest Services.

——— 1977. Contributions to Papuasian ornithology VI. Survey of the birds of southern Bougainville Island, Papua New Guinea. CSIRO Div. Wild. Res. Tech. paper 34.

——— 1978. The identity of five type specimens of New Guinean birds. *Emu* 78: 1-6.

——— 1981. The taxonomy of the RAOU "Interim List of Australian Songbirds". *Austr. Birds* 15: 63-71.

———, B. GLOVER, F. L. KINSKY, S. MARCHANT, A. R. MCGILL, & S. A. PARKER. 1978. Recommended English names for Australian birds. *Emu* 77: 245-313.

———, & I. J. MASON. 1980. Nocturnal Birds of Australia. Melbourne: Lansdowne.

———, ———, & H. B. GILL. 1982. The avifauna of the Australian mangroves—a brief review of composition, structure and origin. *In* Mangrove Ecosystems in Australia. Structure, Function and Management. B. F. Clough (ed.). Canberra: ANU Press.

———, & J. L. McKEAN. 1973. The species of the genus *Parotia* (Paradisaeidae) and their relationships. *Emu* 73: 145-156.

———, & ——— 1976. The relationships of some monotypic genera of Australian oscines. Pp. 530-541, *in* H. J. Frith, & J. H. Calaby (eds.), Proc. 16th Int. Orn. Congr. Canberra: Austr. Academy of Sci.

———, G. A. SMITH, I. J. MASON, & R. G. WEATHERLY. 1979. Relationships and speciation in the Australian corellas (Psittacidae). *Bull. Brit. Orn. Cl.* 99: 128-137.

———, & R. G. WEATHERLY. 1982a. The Fairy-wrens, a Monograph of the Maluridae. Melbourne: Lansdowne Editions.

———, & ——— 1982b [1983]. Campbell's Fairy-wren *Malurus campbelli*, a new species from New Guinea. *Emu* 82: 308-309.

SIBLEY, C. G. 1970. A comparative study of egg-white proteins of passerine birds. *Bull. Peabody Mus. nat. Hist. Yale Univ.* 32: 1-131.

SIBLEY, C. G., & J. E. ALHQUIST. 1982a. The relationships of the Australo-Papuan scrub-robins *Drymodes* as indicated by DNA-DNA hydridization. *Emu* 82: 101-105.

———, & ——— 1982b. The relationships of the Australo-Papuan sittellas *Daphoenositta* as indicated by DNA-DNA hybridization. *Emu* 82: 173-176.

———, & ——— 1982c. The relationships of the Australasian whistlers *Pachycephala* as indicated by DNA-DNA hybridization. *Emu* 82: 199-202.

———, & ——— 1982d. The relationships of the Australo-Papuan fairy-wrens as indicated by DNA-DNA hybridization. *Emu* 82: 251-255.

———, & ——— 1984. The relationships of the genus *Peltops. Emu* 84: 181-183.

———, & ——— In press. The phylogeny and classification of passerine birds, based on comparisons of the genetic material, DNA. Moscow: Proc. 18th Int. Orn. Congr.

———, R. SCHODDE, & J. E. AHLQUIST. 1984. The relationships of the treecreepers Climacteridae as indicated by DNA-DNA hybridization. *Emu* 84: 236-241.

SLATER, P. 1970. A Field Guide to Australian Birds, Volume 1. Non-Passerines, Adelaide: Rigby.

——— 1974. A Field Guide to Australian Birds, Volume 2. Passerines. Adelaide: Rigby.

SMITH, G. A. 1975. Systematics of parrots. *Ibis* 117: 18-68.

SMYTHIES, B. E. 1981. The Birds of Borneo. 3rd ed. Kuala Lumpur: Malayan Nature Society.

SOMADIKARTA, S. 1967. A recharacterization of *Collocalia papuensis* Rand, the Three-toed Swiftlet. Proc. U.S. Nat. Mus. 3629.

STEINBACHER, J. 1979. Threskiornithidae. Pp. 253-268, *in* E. Mayr, & G. W. Cottrell (eds.), Checklist of Birds of the World, Volume 1 (2nd ed.). Cambridge: Mus. comp. Zool.

STOLL, N. R. , R. Ph. DOLLFUS, J. FOREST, N. D. RILEY, C. W. SABROSKY, C. W. WRIGHT, & R. V. MELVILLE. 1963. International Code of Zoological Nomenclature. London: Inter. Trust for Zool. Nomenclature.

STONOR, C. R. 1937. On the systematic position of Ptilonorhynchidae. *Proc. Zool. Soc. Lond.* 107: 475-490.

STORER, R. W. 1979. Podicipedidae. Pp. 140-155, *in* E. Mayr, & G. W. Cottrell (eds.), Checklist of Birds of the World, Volume 1 (2nd ed.). Cambridge: Mus. comp. Zool.

STORR, G. M. 1973. List of Queensland Birds. Spec. Publ. West Austr. Mus. 5.

STRESEMANN, E. 1941. *Calidris ferruginea* (Pontoppidan) statt *Calidris testacea* (Pallas). *Ornith., Monatsber.* 49: 21.

———, & D. AMADON. Falconiformes. Pp. 271-425, *in* E. Mayr, & G. W. Cottrell (eds.), Checklist of Birds of the World, Volume 1 (2nd ed.). Cambridge: Mus. comp. Zool.

STRONOCH, N. R. H. 1980. Chestnut Teal *Anas castanea* at Bensbach (Western Province), a new species for the New Guinea region. *PNG Bird Soc. Newsl.* 173-173: 3.

——— 1981a. Grey Falcon at Bensbach, Western Province, a new species for the New Guinea region. *PNG Bird Soc. Newsl.* 178-179: 5.

——— 1981b. Pallid Cuckoo *Cuculus pallidus*, a new species for Papua New Guinea. *PNG Bird Soc. Newsl.* 181-182: 23.

THOMPSON, H. A. F. 1979. First Australasian record of Sand Martin *Riparia riparia*. *PNG Bird Soc. Newsl.* 151: 6-7.

TUBB, J. A. 1945. Field notes on some New Guinea birds. *Emu* 44: 249-273.

VAN TETS, G. F. 1977. [Review of] Systematics and evolutionary relationships among the herons (Ardeidae) by R. B. Payne and C. J. Risley. *Emu* 77: 156-157.

VAURIE, C. 1962. Dicruridae. Pp. 137-157, *in* E. Mayr, & J. C. Greenway (eds.), Checklist of Birds of the World, Volume 15. Cambridge: Mus. comp. Zool.

——— 1965. Birds of the Palearctic Fauna. Non-Passeriformes. London: Witherby.

———, C. M. N. WHITE, E. MAYR, & J. C. GREENWAY. 1960. Motacillidae. Pp. 129-167, *in* J. L. Peters (ed.), Checklist of Birds of the World, Volume 9. Cambridge: Mus. comp. Zool.

VOOUS, K. 1977. List of Recent Holarctic Bird Species. London: BOU.

WALKINSHAW, L. 1973. Cranes of the World. New York: Winchester.

WETMORE, A. 1930. A systematic classification for the birds of the world. *Proc. U.S. Natn. Mus.* 76(24): 1-8.

WETMORE, A. 1957. The classification of the oscine Passeriformes. *Condor* 59: 207-209.

———1960. A classification for the birds of the world. *Smithson. Misc. Collect.* 139(11): 1-37.

WHEELER, J. R. 1969. New bird species for PNG. *Geelong Nat.* 3: 125.

WHITE, C. M. N., & M. D. BRUCE. 1985. Birds of Wallacea (Sulawesi, Moluccas and Lesser Sunda Islands). BOU checklist series.

WOOD, V. J. 1970. The breeding and distribution of *Elanus caeruleus* in New Guinea. *Sunbird* 1: 48-55.

ZUSI, R. L., & BENTZ, G. D. 1982. Variation of a muscle in hummingbirds and swifts and its systematic implications. *Proc. Biol. Soc. Wash.* 95: 412-420.

NOMENCLATURAL INDEX

All names mentioned in the Annotated List are included in the following Index. Note that the numerical references are *not* to page number, but instead refer to each *species number*. For group names, the species number for the first species of each is listed.

ADVICE TO CONTRIBUTORS

RAOU MONOGRAPHS is a vehicle for the publication of ornithological studies in the Australasian region. It is published by the Royal Australasian Ornithologists Union and is designed specially for monographic works too long to warrant inclusion in the Union's quarterly journal, *The Emu*. Papers will be published in the order of receipt but priority will be given to those in which the author(s) can contribute to the costs of publication. Papers on any aspect of ornithology in the Australasian region are invited and should be sent to the following address:

Mr Leo Joseph
Editor, Austral. Orn. Monographs
RAOU
21 Gladstone Street
Moonee Ponds
VICTORIA
AUSTRALIA 3039.

Three copies of the complete manuscript should be sent to the Editor. A general guide to style and layout will be found in *The Emu*. Authors should consult the Editor when in doubt about current usage and style.

All material in the manuscript, including the abstract (not to exceed three pages), table of contents if applicable, text, tables, table headings, figure legends, footnotes and references must be typewritten, double spaced and on A4 (215 × 290 mm) paper. Scientific names of genera, species and subspecies should be underlined. All vernacular names of plants and animals should be accompanied by a scientific name the first time they are used.

Each figure should be on a separate sheet numbered with an Arabic numeral. All maps, graphs, charts, drawings and photographs should be numbered as figures in the sequence in which they are referred to in the text. Illustrations should be sufficiently large to allow reduction to the size of a page of print or part thereof.

Table headings should be concise, but should provide an adequate description of the data presented.

Literature should be cited according to the style of *The Emu*. It is the responsibility of the author to check carefully all references against the original sources. The Editor does not accept that responsibility.

Authors will be supplied page proofs and they must return proofs promptly to the Editor. Orders for extra copies of each paper must be sent to the Editor when proofs are returned. Extra copies will be supplied to the authors and sponsors at cost.